Advanced Materials for Energy and Environmental Applications

Editor

Sunghoon Park

MDPI • Basel • Beijing • Wuhan • Barcelona • Belgrade • Manchester • Tokyo • Cluj • Tianjin

Editor
Sunghoon Park
Soongsil University
Korea

Editorial Office
MDPI
St. Alban-Anlage 66
4052 Basel, Switzerland

This is a reprint of articles from the Special Issue published online in the open access journal *Materials* (ISSN 1996-1944) (available at: https://www.mdpi.com/journal/materials/special_issues/materials_energy_environmental_applications).

For citation purposes, cite each article independently as indicated on the article page online and as indicated below:

LastName, A.A.; LastName, B.B.; LastName, C.C. Article Title. *Journal Name* **Year**, *Article Number*, Page Range.

ISBN 978-3-03943-072-7 (Hbk)
ISBN 978-3-03943-073-4 (PDF)

Contents

About the Editor

Sunghoon Park, Ph.D.
Assistant Professor, Department of Mechanical Engineering,
Soongsil University, Seoul 156-743, Korea
E-mail: leopark@ssu.ac.kr, leo2328@gmail.com
Webpage: https://sites.google.com/site/surfcomplab/paper

EXPERIENCE
2015–Current: Assistant Professor in Mechanical Engineering, Soongsil University
(Soongsil Fellowship Professor)
2009–2015: Samsung Advanced Institute of Technology (SAIT),
Senior Engineer in the Material Center

EDUCATION
2009 Ph.D. in Materials Science and Engineering, University of California, San Diego (UCSD),
Department of Mechanical and Aerospace Engineering (MAE).

Preface to "Advanced Materials for Energy and Environmental Applications"

Advanced materials for energy and environmental applications (such as rapid heating, anti-fouling/anti-virus surface, chemical sensor, textile/stretchable sensor, fuel cell, and lithium-ion batteries) have been extensively investigated in the academic and industrial fields. The advent of carbon-based nano-materials (carbon nanotubes, graphene, and carbon black) and inorganic nano-materials (Ag wire/particles, Cu mesh, and transition metal dichalcogenide) has accelerated research interest in energy and environmental applications.

This book is focused on the emerging concept and improvement of energy and environmental basic research, as well as in the characterization and analysis of novel energy and environmental base materials. The contents of the book are as below:

- Theoretical and experimental studies on advanced conducting nanocomposites;
- Electrical properties of nanocomposites under various conditions (dynamic mode, aspect ratio, alignment, and contents) and its applications;
- Advanced material for energy applications;
- Analysis and materials for environmental applications.

Sunghoon Park
Editor

materials

MDPI

Article

Numerical Characterization for Electrical Conductivity of Two-Dimensional Nanocomposite Systems with Conducting Fiber Fillers

Jungmin Lee [1], Yesol Yun [2], Sang Hyun Lee [2,*] and Jinyoung Hwang [1,*]

1 School of Electronics and Information Engineering, Korea Aerospace University, Goyang-si 10540, Korea; jeungmin95@kau.ac.kr

2 School of Electrical Engineering, Korea University, Seoul 02841, Korea; yesolar@korea.ac.kr

* Correspondence: sanghyunlee@korea.ac.kr (S.H.L.); jinhwang@kau.ac.kr (J.H.); Tel.: +82-2-300-0422 (J.H.)

Received: 4 May 2020; Accepted: 21 May 2020; Published: 24 May 2020

Abstract: Hybrid nanotube composite systems with two different types of fillers attract considerable attention in several applications. The incorporation of secondary fillers exhibits conflicting behaviors of the electrical conductivity, which either increases or decreases according to the dimension of secondary fillers. This paper addresses quantitative models to predict the electrical performance in the configuration of two dimensional systems with one-dimensional secondary fillers. To characterize these properties, Monte Carlo simulations are conducted for percolating networks with a realistic model with the consideration of the resistance of conducting NWs, which conventional computational approaches mostly lack from the common assumption of zero-resistance or perfect conducting NWs. The simulation results with nonperfect conductor NWs are compared with the previous results of perfect conductors. The variation of the electrical conductivity reduces with the consideration of the resistance as compared to the cases with perfect conducting fillers, where the overall electrical conductivity solely originates from the contact resistance caused by tunneling effects between NWs. In addition, it is observed that the resistance associated with the case of invariant conductivity with respect to the dimension of the secondary fillers increases, resulting in the need for secondary fillers with the increased scale to achieve the same electrical performance. The results offer useful design guidelines for the use of a two-dimensional percolation network for flexible conducting electrodes.

Keywords: nano-composites; Monte Carlo simulation; percolation networks; electrical conductivity

1. Introduction

Conducting polymer composites have become essential components for various modern electronic applications including wearable sensors, flexible displays, batteries, and solar cells, etc. [1–4]. In particular, CNT-based composites have recently been studied for specific improved properties in those applications [5–7]. The percolation networks of one-dimensional nanostructures, two-dimensional materials, such as graphene, and conducting polymers, such as PEDOT:PSS and polyaniline, are considered for their key enablers. In particular, percolation networks of one-dimensional nanowires (NWs) have been extensively studied for applications of flexible conducting polymers. Ultra-high aspect ratio one-dimensional fillers of highly conductive materials, such as carbon nanotube (CNT), gold (Au), silver (Ag), copper (Cu), and their alloys, are desired to achieve the outstanding structural and electrical performance at low coverage. Among those options, CNT-based polymer composites have been intensively investigated for very attractive options, in particular, of their extremely large aspect ratio and expanded surface area. Furthermore, high-quality NWs can be synthesized via low-cost and scalable wet-based process, and a random NW network can be fabricated by coating a dispersed NW solution on a flexible substrate using a slot-die method, bar-coating, and spin coating, and spray coating process [2,4]. Futhermore, a number of studies have addressed the improvement of the electrical performance of the NW network. CNTs/Cu composites show fascinating electrical and thermal conductivities with a larger value of thermal expansion (CTE) coefficient than copper [8]. CNTs based composites also have been utilized in the design of microscopic flexible electronics, such as high-performance Li-S batteries [9], Li-ion batteries [10], and capacitive strain sensors for human motion detection [11,12]. Furthermore, nitrogen-doped CNT-based composites are composite bifunctional catalysts for improving the performance of the rechargeable zinc-air battery [13]. For device applications, this system can be applied to an electromagnetic shield for light-weight and flexibility, and an alternative for metal conductors that are erosive, heavy-weight, costly.

For additional improvement of the performance, the incorporation of particulate fillers can be considered to form the excluded volume that effectively leads to a segregated CNT network [14–18]. In consequence, the probability of CNTs forming a conducting network increases in the segregated composite system, indicating that much lower CNT loading suffices to reach a certain degree of the electrical conductivity [15]. Thus, as compared to CNT composites made only of CNT fillers, the segregated composite system could effectively relieve the cost issue of CNTs. From a mechanical point of view, dramatic strengthening and toughening, which are essential requisites for practical applications, can be achieved by particulate fillers, such as silica, $CaCO_3$, and $BaSO_4$. It has an effect on creating electrical junctions between CNTs. The incorporation of fillers into CNT-filled polymers can reduce the space available for CNTs to form conductive networks and could increase the conductivity because CNTs cannot diffuse into the fillers which are solid particles [19–22]. Therefore, CNT-based polymer composites, which are comprised of CNTs and micro-scale or nano-scale secondary fillers, change the electrical conductivity due to the side effect of secondary particulate fillers. Micro-scale fillers increase the electrical conductivity for the excluded volume, while nano-scale secondary fillers decrease the electrical conductivity.

This work addresses the computational characterization of the effect of the dimensional properties of secondary particulate fillers on the electrical conductivity of CNT polymer composites. In the previous research, to clarify the contribution of the size effect of secondary particulate fillers on the electrical conductivity of the nanocomposite network, the perfect conductor is assumed for the electrical property of the CNT filler, i.e., the resistivity of the CNT is set to zero to conduct Monte Carlo (MC) simulation for the prediction of percolating behaviors of the CNT network in CNT/silica composites [23]. Thus, the contact resistance of the CNT network is solely a source of the network sheet resistance. Therefore, this work proposes to evaluate the electrical conductivity of the percolating nanocomposite network with the consideration of the resistivity of CNT materials, i.e., nonperfect conducting materials are used for one-dimensional conducting fillers. In addition, we provide the refined interpretation of the change in the electrical conductivity of the CNT composites with respect to the size of particulate fillers via the combination of Swiss cheese models and the underlying percolation theory with nonperfect conducting fillers. The results allow to predict the electrical performance of nanocomposites in a fast and simplified manner as compared to conventional methods for the same purposes.

The conflicting electrical behaviors caused by the addition of secondary particulate fillers can be briefly described with Swiss cheese model. A random configuration of secondary particulate fillers on a polymer matrix can be modeled in a shape of Swiss cheese with many pores. A pore represents a single secondary filler which excludes conducting fibers, and the volume of "cheese" is the space that conducting fibers can occupy. A network of conducting fillers presents the collection of all conducting paths in the polymer matrix with particulate fillers. The CNT network is separated from the conducting fiber composite to evaluate the effective conductance of conducting paths. The local conductance of a point within the volume of Swiss cheese model depends on the density of CNT fibers placed around the corresponding region. To assess the degree and formation of an electrical network and to calculate the total conductance, an available configuration of the effective conducting network is obtained from the joint consideration of the conducting path and geometry of the particulate fillers. Thus, the total conductance can be evaluated by considering the continuum percolation property of the effective network. The transition of the electrical conductivity is observed in accordance with the excluded volume caused by secondary particles, i.e., the excluded volume improves or prevents the conductivity of CNT composites depending on the particle size. The network morphology of CNT composites depends on the size of the particulate fillers, resulting in the variation of the electrical conductivity. From this model, a numerical characterization of conflicting conducting behaviors is carried out based on a unified framework. Since most of available NW composite networks are created with conducting but nonzero resistance fillers, understanding of the impact on the electrical properties of nonzero resistance fillers is necessary. This work develops quantitative models of the electrical conductivity of two-dimensional NW composite networks with one-dimensional nonperfect conducting fillers via MC based computation. Based on computational models, the change of the network sheet resistance involved with the consideration of the NW resistivity is investigated with respect to various design factors for performance improvement. The results are compared with the perfect conducting NW percolating network. According to the results, the network conductance exhibits a relatively small change with the change of the filler size variation, resulting in the change of linear slopes, which suggests the prediction of the network electrical conductivity with an enhanced accuracy. The corresponding performance is also demonstrated in a computational way.

2. Method

The MC simulation is conducted to identify percolating behaviors of the CNT network in CNT/silica composites under various size and concentration conditions of composite components. In order to predict the sheet conductance (σ) of a one-dimensional NW percolation network, a random instance of the NW network is generated in a square domain by placing NWs and particulate fillers. A sample square domain of 25 μm × 25 μm is considered to obtain the CNT network. A single flexible CNT wire in the composites is modeled as a long thin cylindrical object with many joints to reflect the bending of the wire, while silica particles are represented in a circular shape with the diameter equal to an average size of silica particles.

Circular particles are scattered in the two-dimensional domain via random choices of central coordinates of nonoverlapping individual particles. For an accuracy of the quantitative analysis with the total volume of the particulate fillers, all silica particles lie nonoverlapping on the domain. To this end, the center coordinates of individual circular objects are chosen uniformly at random inside the domain. Subsequently, the fillers are placed sequentially so that the distances from the chosen center position to the center positions of all pre-existing spheres are apart by larger than the diameter. The number of particles to be placed in the domain is determined from the ratio of the silica volume to the domain volume.

Upon the completion of the placement of all nonoverlapping particulate fillers, a CNT network is created on the domain with multiple random instances of a flexible CNT wire. A NW is placed uniformly on the simulation domain so that it does not penetrate pre-located particulate fillers but bends to circumvent them. Figure 1 illustrates the generation of a single bendable conducting CNT wire. The flexibility of the NW is reflected by modeling it geometrically as a series of small line segments contiguously connected with joints that are allowed to bend freely within a certain degree, which is chosen as 120° from experimental results [24], such that the width and length of a single segment equal to the diameter and length of a cylindrical NW, respectively. To create a single flexible NW on the simulation domain, constituting line segments are placed in sequence from one end to the other. A random choice of the position of the first line segment is examined if it is placed in the domain that is not occupied by any spherical filler. If the position of the first segment proves valid, a random direction which the NW stretches toward is subsequently chosen. The next line segments are repeatedly placed in a straight line unless touching pre-occupying spheres. If the candidate position of the next segment is predicted to be inside a spherical filler, the line segment bends to change the direction toward a tangential direction of the spherical shape at the contact point to avoid an undesired penetration so that a flexible wire smoothly circumvents the sphere to stretch forward. The length of line segments forming a single flexible wire is set the same as the width of the wire, which facilitates the test for the contact of two flexible wires. Thus, the line segment can be viewed as a square shape with the identical height and width. The test for the contact is sufficient to examine if another line segment from a different NW lies within a spherical ball of radius equal to the length of the line segment. The range of the bending angle of the joint between two consecutive line segments is limited to 120°, which is determined by the physical property of a CNT where the carbon tube structure can be preserved with very large bending angle without atomic destruction for its perfect hexagonal structure of a CNT NW. If a CNT/silica composite network is constructed, the connection status of NWs in the CNT network is examined by checking the contact of line segments from all other NWs in the domain. The connectivity between two wires is determined by calculating all possible pairwise distances between line segments belonging to respective wires. If the shortest distance between the two segments is less than the distance of the tunneling effect, the NWs are considered to touch each other. To consider the NW resistance in the network, a novel technique is developed forsimulation. The NW resistance comes into play in the network if a single NW forms a conducting path of the percolating network. This happens when the NW has contacts with at least two different NWs. Since the resistance is proportional to the length of the resistor, the length of the portion in between two contact points of the single NW determines the overall resistance straightforwardly. An efficient way of calculating the distance between two contact points is to count the number of line segments connecting two contact points. While placing individual line segments to create a single NW, the number of line segments is counted to obtain the desired length of the NW. Once a new line segment is found to touch another line segment from a different NW, the index of the corresponding NW is available from the count of line segments. Since the count of line segments between two adjacent contact points, which are obtained by the difference between indices of contact line segments, is proportional to the distance between two contact points. The count of line segments n is multiplied by the resistivity of NW ρ and the length of the line segment w, which is chosen identical to the width of the NW, to calculate an equivalent resistance given by $n\rho w$. Thus, equivalent resistances vary with the distance between contact points. If a single NW has more than two contact points with adjacent NWs, all pairs of two consecutive contact points act as a serial connection of equivalent NW resistors with contact resistance bypasses connected at each junction, as illustrated in Figure 1.

The connectivity information of an individual NW is collected upon the completion of its placement to form an adjacent matrix representing the relationship among NWs. Based on this information, a clustering analysis for the CNT network is conducted via union-find algorithm for exploring percolating clusters that traverse across the square domain. An individual cluster is expanded gradually toward the right end by including NWs in contact, beginning with the leftmost NWs touching the left end of the simulation domain. If two paths share a conducting NW, those paths are agglomerated as a cluster to proceed in a single path. If a conducting path touches the right end of the domain, the corresponding NW agglomeration is found as a conducting cluster. The NWs that are not contained in the conducting cluster are discarded to construct an adjacency matrix. The adjacency matrix is constructed in a square matrix with each column and row associated with an individual contact point and the corresponding entry equal to either of contact and NW resistance.

Using the adjacency matrix, the overall resistance of the CNT network is calculated based on Kirchhoff's current law (KCL) derived for the network with a 1V voltage source applied at both ends of the domain by considering an equivalent circuit network with two types of the resistance arising from inherent NW resistance and contact resistance between two NWs. KCL is applied at all contact points in the cluster to formulate a system of linear equations with respect to the voltage drop at every contact point in the conducting path. The solution of the system of linear equations, which can be obtained using linear algebra software packages. The total current flowing across the square domain is calculated from the solution. From the assumption of the 1V voltage source, the obtained total current corresponds to the overall conductance σ of the CNT network. The parameters of the NW considered in the simulation are set according to experimentally measured properties of currently available NWs. A NW is geometrically modeled as a cylinder with a diameter of 15 nm, the length of 5 μm, and the resistance of 2×10^{-2} Ω [25]. The contact resistance for two NWs in contact is set to 500 Ω, considering current state-of-the-arts in fabrication technology [26]. The contact resistance is constant over the NW network since the tunneling resistance in the two-dimensional NW network does not undergo a large variation in the distance between NWs. The above procedure is repeated to obtain 500 independent random instances, and an ensemble average of the resistance is calculated based on the set of tested samples between 30th percentile and 70th percentile. The simulation is conducted over different configurations of silica content, silica size, CNT content, and CNT dimension. For massive simulation, the simulation is implemented in parallel processing package such MATLAB computing toolbox.

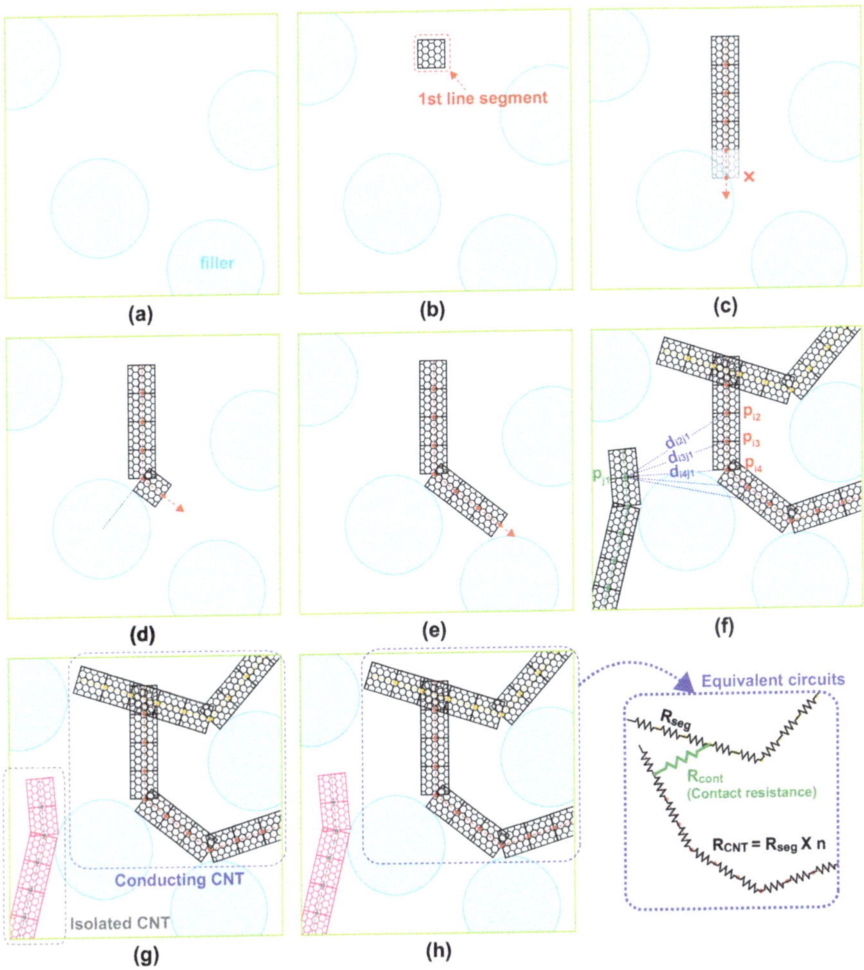

Figure 1. Generation of an instance of CNT/silica composite in Monte Carlo simulation: (**a**) placement of particulate fillers, (**b–e**) creation of a single bending conducting filler, (**f**) examination of the connection status of NWs, (**g**) clustering analysis of the NW network, and (**h**) equivalent circuits of the NW network for KCL application.

3. Results

The size effect of secondary particulate fillers on the electrical resistivity of the nanocomposites can be observed with random instances of the CNT/silica composites generated from the simulation presented in Figure 2. Figure 2a–c are percolating CNT networks of the same coverages of CNT NWs and particulate fillers for three cases of no secondary filler, micro-scale filler, and nano-scale fillers, respectively. A CNT network forming conducting paths is represented in black, while the remaining NWs isolated from the percolation network are represented in purple. Figure 2a shows a random network of CNTs with the diameter of 15 nm and the length of 5 μm in the simulation domain of 25 μm × 25 μm. In the simulation setup, the CNT content of 4% is dispersed in an epoxy resin. Since the percolation occurs in this configuration of the CNT network, all NWs are included in Conducting networks. By contrast, Figure 2b,c illustrate representative random instances of CNT/silica composites generated with the same loading amount of CNTs along with the micro-size (the diameter of 4 μm) and the nano-size (the diameter of 50 nm) particulate fillers, respectively, at the silica content of 20 wt%. In comparison with the case without secondary fillers, the micro-scale silica CNT composite network contains several dense CNT clusters formed by the excluded volume arising from the micro-size silica fillers, which is consistent with the results of existing studies about the excluded volume theory. Topological changes of the network, such as dense CNT clusters, lead to the enhancement in the chance that NWs in the domain CNTs are interconnected to form a large number of contacts. The total number of contact points in den CNT clusters forming major conducting paths in the percolating network increases from 502 to 2599 after incorporating micro-scale silica fillers. The increase in the number of contact points in the network evidences the improvement of the network electrical conductivity over nanocomposite systems without secondary fillers.

On the other hand, nano-scale secondary fillers prevent NWs from stretching toward the right side of the domain, and densely dispersed nano-sized particulate fillers cause the resulting system an abrupt increase with severe NW kinks. As shown in Figure 2c, the same loading amount of NWs covers only a fraction of the conducting network. The resulting topology change disrupts the formation of a conducting CNT network. The total number of interconnecting junctions in active percolating clusters decreases from 502 to 269. These results are consistent with the resistivity increase of the CNT/nano-size silica composite measured empirically. To see this in more detail, Figure 2d zooms in a boundary of conducting and nonconducting network regions in Figure 2c. In s micro-scale filler case, all regions unoccupied by two micro-scale particulate fillers between them, referred to as necks in percolation theory, have at least one NW passing. Regarding the percolation theory, a neck acts as a conducting path in the percolating network, and thus all necks participate in the conducting network. By contrast, only a few necks between nano-scale fillers have NWs passing through them, implying that a small number of necks act as a conducting path in the CNT network. Furthermore, the NW kinks incurred by the blockage of nano-scale fillers reduces the effective length of conducting NWs. Therefore, there is little chance that all NWs forms a conducting CNT network, and only a fraction of the two-dimensional domain has a conducting property, resulting in a large network sheet resistance of the network.

In order to assess a quantitative impact of the filler size, the electrical conductivity is monitored with varying secondary particulate filler contents with the CNT content fixed to 4%, which corresponds to the above electrical percolation threshold of the nanocomposite system. Figure 3 shows the change in the network sheet resistance for two cases of perfect conducting NWs and nonzero NW resistance. The change in the sheet resistance is represented by the ratio of the sheet conductivity without (σ_0) and with (σ) the secondary particulate fillers, i.e., σ/σ_0. Figure 3a shows the change of the conductivity of percolating networks consisting of perfect conducting NWs with zero resistance. In case of micro-scale particulate fillers, the increase in the micro-size filler loading content increases the electrical conductivity of the CNT network. According to the excluded volume theory, the improvement of the electrical conductivity relies on the excluded volume of secondary fillers driving concentrated NWs to create a dense CNT network. On the other hand, the incorporation of nano-scale silica fillers decreases the electrical conductivity as the silica filler content increases. This disruption of the CNT network degradation is attributed to the distortion of CNTs caused by dispersed nano-sized silica particles, and these conflicting behaviors caused by the excluded volume have been first explained with Voronoi geometry in [23]. Furthermore, the results reveal the existence of the secondary filler configuration that results in the invariance of the network conductivity with the change of the secondary filler coverage, corresponding to nanoparticle fillers with the diameter of 300 nm. Since such a point exists in any hybrid nanocomposite system consisting of multiple types of filler, this can be viewed as a criterion for comparison of the electrical conductivity performance.

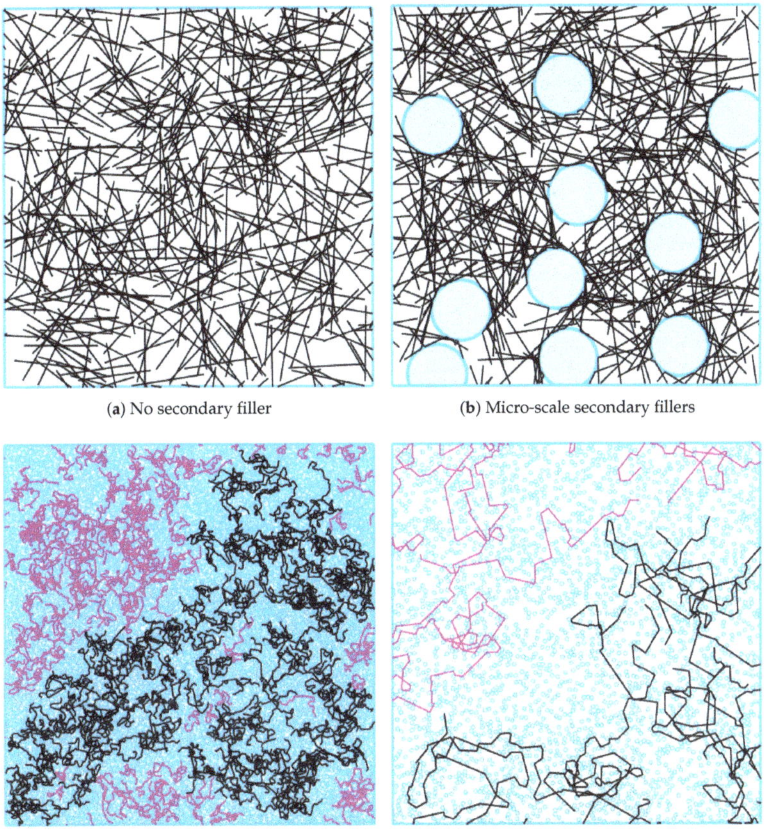

(**a**) No secondary filler (**b**) Micro-scale secondary fillers

(**c**) Nano-scale secondary fillers (**d**) Boundary of conducting networks

Figure 2. Random percolating networks for various cases.

Figure 3b shows the change of the conductivity of the CNT network with nonzero resistance nonperfect conducting NWs. This configuration shows similar trends of the change in the electrical conductivity, i.e., the conductivity ratio σ/σ_0 increases with increasing particulate filler sizes, whereas it reduces with the decreasing filler size with nano-scale particulate fillers, which are comparable to the corresponding conductivity change in the experimental data [23,27–29]. It is observed, however, that the slope of the conductivity change diminishes. The slope of the new system reflects a linear trend in conductance change with respect to the silica content in a log scale, as shown in experimental data [23]. Thus, the overall network conductivity is less sensitive to the contribution of the contact resistance. Furthermore, the size of particulate fillers leading to the conductivity invariance rises from 300 nm to 1.5 μm, which amounts to five times larger than the previous case. Thus, the network topology associated with the new case contains the same loading amount of NWs and secondary fillers but of a large scale. This is naturally expected since the consideration of additional resistance causes the increase of the overall resistance, leading to the decreased current. In order to compensate for the decreased current, additional conducting paths are necessary and can be achieved by allowing additional dense CNT clusters in the conducting network, which can be achieved with the increase in the secondary filler size. In fact, the value of overall network conductivity without secondary fillers σ_0 is increased.

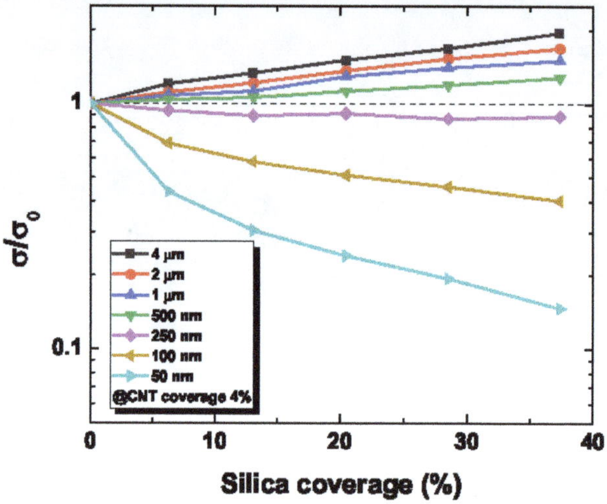

(**a**) Perfect conducting fillers

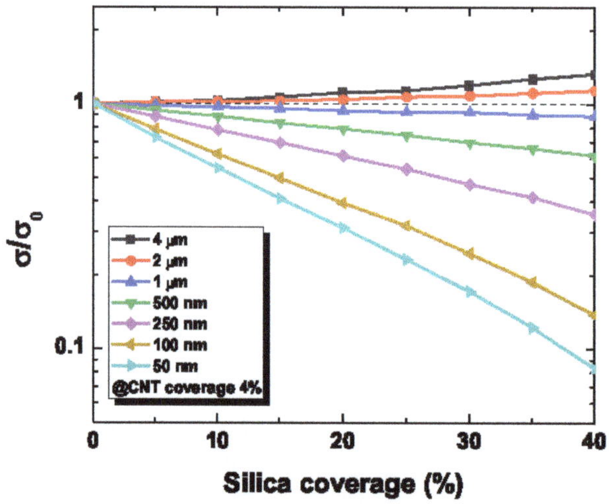

(**b**) nonperfect conducting fillers

Figure 3. Normalized conductance changes for perfect and nonperfect conducting fillers with respect to silica coverage.

Figure 4 plots the changes in the sheet conductance of the percolating NW networks with respect to the value of the NW resistance denoted by R_{CNT} in the secondary filler configuration of the diameter of 4 μm and content coverage of 40%. Random instances of the NW network considered in the simulation are basically identical except for the value of the NW resistance. In order to observe the degree of the difference, the normalized conductance change, defined as σ/σ_0, is measured to compare the average resistance of the composite system with and without silica fillers for increasing CNT contents under the fixed secondary silica content. The particulate filler effect on the conductivity of the composite system is maximized (or the most deviated from the original value) in low CNT loading regimes, while particulate fillers do not exhibit a significant impact in high CNT loading contents consistently for all cases of the NW resistance. The comparison among NW resistances indicates that the conductivity change is less sensitive for a large value of the NW resistance. The NW resistance and contact resistance contribute to the overall network resistance additively. In addition, the identical topology of the CNT networks leads to identical contact resistances for different configurations of NW resistance. This implies that the resulting normalized conductance change, defined as the ratio of the contact resistance contribution to the sum of the contact and NW resistance contributions, has an inversely proportional relationship with the value of NW resistance. This observation also reveals that the case with large contact resistance has the closest limit of the normalized conductance change toward one since the contribution of the NW resistance is minimized.

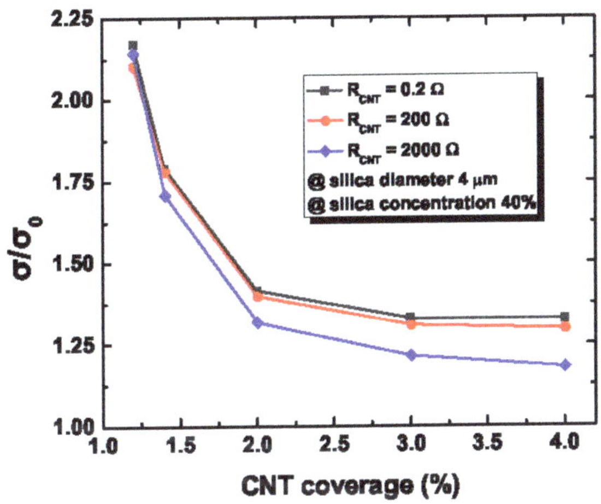

Figure 4. Normalized conductance changes with respect to NW coverage.

4. Discussion

This section provides the discussion about qualitative interpretations and underlying intuitions of quantitative results obtained from the MC simulation. A unified formalism is addressed for the analysis of the normalized conductivity change observed concerning the size of secondary particulate fillers. It can be established via the combination of discrete percolation theory and Swiss cheese model. Hybrid NW systems with particular secondary fillers can be viewed as a novel extension of Swiss cheese model. Swiss cheese model is indeed a porous medium percolation model where insulating particles occupy pores, while continuum such as fluid is allowed to flow for percolation. However, this nanocomposite system is nontrivial in that NWs which act as conducting paths which electrical current can flow through are overlaid in Swiss cheese model originally intended for the continuum percolation. In ordinary Swiss cheese model, the conductivity of a conducting path is characterized by the width of a neck, which is defined as a narrow path between surfaces of two neighboring spherical fillers. The conducting property of this system differs from ordinary Swiss cheese model, where the total conductance depends on the narrowest neck width since only a neck having conducting NWs can act as a valid conducting path. The density of NW clusters passing across the neck determines the electrical conductance of the associated neck. Thus, only a neck with nonzero conductance is a valid conducting path of the percolating network. For example, if a small number of NWs exist in the nanocomposite system, most necks do not contribute to the percolating network, resulting in low network conductance.

This framework is extended to the analysis of the nanocomposite system with micro- and nano-scale fillers. The nanocomposite system's total conduct depends on two network parameters of the numbers of parallel conducting paths and the conductance of individual paths. A conducting path is formed by interconnecting active necks that contain conducting NWs so that it traverses the simulation domain, and the number of such conducting paths connected in parallel determines the amount of the total electrical current flowing across the domain. In addition, a conducting NW can be decomposed into two components in a horizontal direction toward the right end of the domain and its perpendicular vertical direction. The effective length of a conducting NW is defined as the length of the horizontal component of the conducting NW. A large effective length of conducting NW leads to a small number of contacts contained in a conducting path traversing the domain. The contact resistance of the resulting conducting path becomes small. On the other hand, the overall NW resistance does not differ among conducting paths since its value is proportional to the length of the conducting path, which is mostly similar to the length of the domain. Thus, the contact resistance has a dominant contribution to the overall conductance of a conducting path. The addition of micro-scale particulate fillers raises the density of conducting NWs along necks, improving the conductance of conducting paths, and increases the effective length of conducting NWs, leading to the reduction of the contact resistance of conducting paths. Therefore, the overall impact on the system is the increase in the conductance of the percolating network. The addition of nano-scale silica fillers, however, causes the decrease of the conductance of the percolating network, since nano-scale fillers block conducting NWs toward the end of the domain and make them twisted. The kinks of the conducting path increase the traversing distance across the domain, resulting in an increase in the NW resistance, and reduce the effective length of conducting NWs so that the contact resistance also grows. Note also that, since the addition of secondary fillers causes the increase and decrease of the network conductance, it is expected to exist a configuration of filler size that makes the conductance almost unchanged and can be estimated from computational results. Moreover, the aspects of the change in the network conductance for increasing silica content are also explained in a consistent way. According to the experiments and simulation results, the network conductance increases as the content of the micro-scale fillers increases. Additional content of the micro-scale filler increases the density of conducting NWs across necks, thereby increasing the number of current paths. For the increase of the nano-scale silica content, the network conductance, in contrast, diminishes as a large density of nano-scale fillers further increases the kinks of the NWs so that the effective length of conducting NWs shrinks. For the CNT/silica composites with silica fillers of a certain value of the diameter, the nanocomposite system's total conductance remains almost fixed with increasing silica content. The effects of the introduction of additional current paths and the average conductance decrement are comparable, canceling out these opposite effects.

The consideration of nonzero NW resistance introduces several new behaviors of the electrical conductivity of the nanocomposite system. Nonzero NW resistance contributes to the total network resistance additively along with the contact resistance. Since the normalized conductance change is the ratio between the contact resistances, the sum of the contact resistance, and the NW resistance, its value is always less than one. The overall NW resistance indeed does not change very much since it depends on the length of the conducting path, which is given by the domain length and is mostly similar among conducting paths. On the other hand, the contact resistance of the nanocomposite system is created by the percolation phenomena. Furthermore, it normally has an exponential relationship with the secondary filler content. Meanwhile, the topology of two cases with and without nonzero NW resistance are identical, and the resulting contact resistance is also identical. Furthermore, the value of the NW resistance is relatively small as compared to the contact resistance, and the resulting normalized conductivity change is strictly less than one. For the increase of the silica content, the contribution of the NW resistance to the overall network conductance remains almost similar since there is little change in the effective length of the domain. By contrast, the contribution of the contact resistance to the network conductance grows since the addition of silica content changes the topology of the conducting network. Therefore, the resulting normalized conductance change becomes relatively small since the numerator and denominator both depend dominantly on the contact resistance.

Note here that the secondary filler configurations leading to invariant conductance are different. The case with nonperfect conducting NW requires a larger size of the particulate filler, as shown in Figure 5. In such a case, the overall conductivity decreases. The density of the conducting path needs to improve is to compensate the conductance decrease to obtain the same electrical performance as the case with perfect conducting NWs. Therefore, secondary fillers with large sizes are necessary, as in Figure 5b.

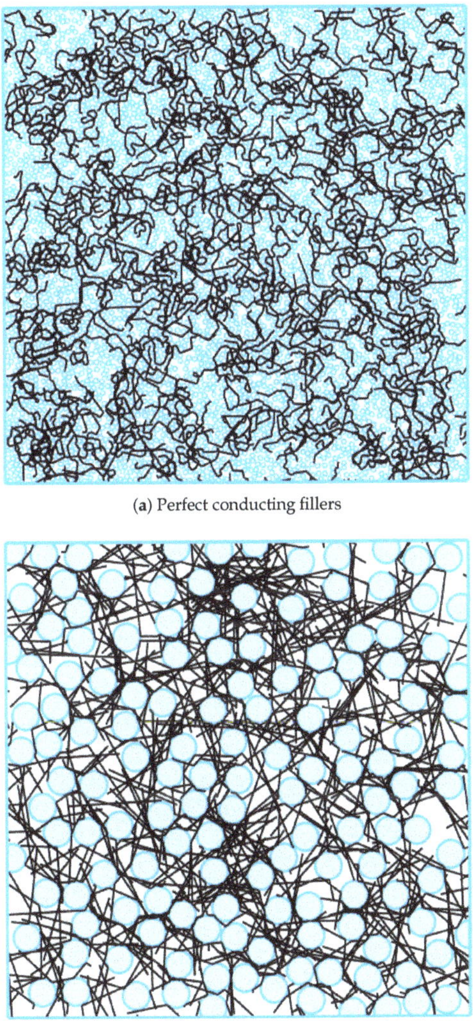

(a) Perfect conducting fillers

(b) Nonzero resistance fillers

Figure 5. Random percolating configurations of conductance invariance with respect to secondary filler sizes.

Finally, the discussion about the conductivity change caused by the CNT content change is ensured. Note that all configurations have the same geometrical topology with the percolating network. The nanocomposite system with larger WN resistance has a larger value of the normalized conductivity change, as in Figure 4. As the CNT content increases in the percolating network, the number of conducting paths increases, and the number of contact resistors that actually contribute to the percolating network decreases since additional CNT creates conducting paths with the reduced number of contact points. Therefore, the conductivity of the percolating network strictly depends on the contact resistance, which becomes gradually diminishing, and its limit approaches unity for increasing CNT content in the nanocomposite system.

In summary, we have shown that nonzero values of the electrical conductivity of NWs affect the conductivity variation of CNT composite networks containing secondary fillers along with the filler size. The overall network resistance increases via the addition of nonzero NW resistance, which has been ignored in the most of previous studies on the electrical property analysis of hybrid nanocomposite systems. In addition, the resulting slope of the change becomes less steep, in a linear change. Furthermore, the secondary filler configuration leading to an invariant network conductivity requires additional current so that it causes the increase with the filler size for nonzero NW resistance. The computational technique developed in this work can handle the electrical properties of comprehensive nanocomposite systems consisting of 1-D conducting wires and insulating particulate fillers such as silica-CNT, ZnO-CNT, and TiO_2-CNT. These achievements shed light on practical design guidelines for the use of the secondary filler in the fabrication of highly-conducting and reinforced CNT composites.

5. Conclusions

This work has developed predictive computational models of the electrical conductivity of the two-dimensional random network with nonzero resistance conducting fillers and suggested strategies designed to enhance the electrical property. The electrical conductance is of a proportional relationship with the size of the secondary filler while showing a consistent decrease with the increasing NW coverage. The consideration of nonzero resistance NWs reduces the impact of the size of the secondary filler on the electrical conductivity of the percolating network, thereby causing linearized changes in the electrical conductivity of the network compared to the case of the perfect conductor. This relationship provides the resistance robustness of the overall network to allow a simplified guideline for the control of electrical properties for hybrid nanocomposite systems. For future research, this work can be extended to study the electrical property of three-dimensional bulk nanocomposites for computational characterization with improved accuracy.

Author Contributions: Conceptualization, S.H.L. and J.H.; methodology, J.L. and J.H.; Software, J.L. and Y.Y.; validation, J.L. and J.H.; formal analysis, J.H. and S.H.L.; investigation, all authors; resources, J.H. and S.H.L.; writing-original draft preparation, S.H.L.; writing-review and editing, J.H. and S.H.L.; visualization, J.L. and Y.Y.; supervision, S.H.L.; project administration, J.H.; funding acquisition, J.H and S.H.L. All authors have read and agreed to the published version of the manuscript.

Funding: This work was supported in part by the National Research Foundation of Korea (NRF) grant funded by the Korea government (MSIT) (No. 2017R1C1B5076588, 2019R1A2C1084855) and in part by a Korea University Grant.

Conflicts of Interest: The authors declare no conflict of interest.

References

1. Heinze, S.; Tersoff, J.; Martel, R.; Derycke, V.; Appenzeller, J.; Avouris, P. Carbon nanotubes as schottky barrier transistors. *Phys. Rev. Lett.* **2002**, *89*, 106801. [CrossRef]
2. Zhang, Q.; Yuan, Z.-Y.; Liu, L.-H.; Zhao, D.; Ji, J.-L.; Zhuo, K.; Zhang, H.-L.; Zhang, W.-D.; Sang, S.-B. Performance of CNTs/GQD-based flexible strain sensors. *J. Nanosci. Nanotechnol.* **2020**, *33*, 1013–1019. [CrossRef] [PubMed]

3. He, H.; Pham-Huy, L.A.; Dramou, P.; Xiao, D.; Zuo, P.; Pham-Huy, C. Carbon nanotubes: Applications in pharmacy and medicine. *BioMed Res. Int.* **2013**, *2013*, 578290. [CrossRef] [PubMed]
4. Han, J.-W.; Kim, B.; Li, J.; Meyyappan, M. Carbon nanotube based humidity sensor on cellulose paper. *J. Phys. Chem. C* **2012**, *116*, 22094–22097. [CrossRef]
5. Ruoff, R.S.; Lorentz, D.C. Mechanical and thermal properties of carbon nanotubes. *Carbon* **1995**, *33*, 925–930. [CrossRef]
6. Salvetat, J.-P.; Bonard, J.-M.; Thomson, N.H.; Kulik, A.J.; Forró, L.; Benoit, W.; Zuppiroli, L. Mechanical properties of carbon nanotubes. *Appl. Phys. A* **1999**, *69*, 255–260. [CrossRef]
7. Che, J.; Tahir Çağın, T.; Goddard III, W.A. Thermal conductivity of carbon nanotubes. *Nanotechnology* **2000**, *11*, 65–69. [CrossRef]
8. Fu, S.; Chen, X.; Liu, P. Preparation of CNTs/Cu composites with good electrical conductivity and excellent mechanical properties. *Mater. Sci. Eng. A* **2020**, *771*, 138656. [CrossRef]
9. Hwang, J.-Y.; Kim, H.-M.; Shin, S.; Sun, Y.-K. Designing a high-performance lithium–sulfur batteries based on layered double hydroxides–carbon nanotubes composite cathode and a dual-functional graphene–polypropylene–Al2O3 separator. *Adv. Funct. Mater.* **2018**, *72*, 121–144. [CrossRef]
10. Liu, X.-M.; dong Huang, Z.; woon Oh, S.; Zhang, B.; Ma, P.C.; Yuen, M.M.; Kim, J.K. Carbon nanotube (CNT)-based composites as electrode material for rechargeable Li-ion batteries: A review. *Compos. Sci. Mater.* **2012**, *28*, 170429. [CrossRef]
11. Jung, S.; Choi, H.W.; Mocanu, F.C.; Shin, D.-W.; Chowdhury, M.F.; Han, S.D.; Suh, Y.-H.; Cho, Y.; Lee, H.; Fan, X.; et al. Modeling electrical percolation to optimize the electromechanical properties of CNT/polymer composites in highly stretchable fiber strain sensors. *Sci. Rep.* **2019**, *9*, 20376. [CrossRef] [PubMed]
12. Cai, L.; Song, L.; Luan, P.; Zhang, Q.; Zhang, N.; Gao, Q.; Zhao, D.; Zhang, X.; Tu, M.; Yang, F.; et al. Super-stretchable, transparent carbon nanotube-based capacitive strain sensors for human motion detection. *Sci. Rep.* **2013**, *3*, 3048. [CrossRef] [PubMed]
13. Chen, Z.; Yu, A.; Ahmed, R.; Wang, H.; Li, H.; Chen, Z. Manganese dioxide nanotube and nitrogen-doped carbon nanotube based composite bifunctional catalyst for rechargeable zinc-air battery. *Electrochim. Acta* **2012**, *69*, 295–300. [CrossRef]
14. Attari, S.G.; Bahrami, A.; Shahna, F.G.; Heidari, M. Single-walled carbon nanotube/silica composites as a novel coating for solid-phase microextraction fiber based on sol-gel technology. *J. Anal. Chem.* **2015**, *70*, 1192–1198. [CrossRef]
15. Peng, H.; Jain, M.; Peterson, D.E.; Zhu, Y.; Jia, Q. Composite carbon nanotube/silica fibers with improved mechanical strengths and electrical conductivities. *Small* **2008**, *4*, 1964–1967. [CrossRef] [PubMed]
16. Song, W.-L.; Cao, M.-S.; Hou, Z.-L.; Fang, X.-Y.; Shi, X.-L.; Yuan, J. High dielectric loss and its monotonic dependence of conducting-dominated multiwalled carbon nanotubes/silica nanocomposite on temperature raging from 373 50 873 K ion X-band. *Appl. Phys. Lett.* **2009**, *94*, 233110. [CrossRef]
17. Kim, D.; Kim, Y.; Choi, K.; Grunlan, J.C.; Yu, C. Improved thermoelectric behavior of nanotube-filled polymer composites with poly(3,4-ethylenedioxythiophene) poly(styrenesulfonate). *ACS Nano* **2010**, *4*, 513–523. [CrossRef]
18. Li, Z.; Nambiar, S.; Zheng, W.; Yeow, J.T.W. PDMS/single-walled carbon nanotube composite for proton radiation shielding in space applications. *Mater. Lett.* **2013**, *108*, 79–83. [CrossRef]
19. Liu, Y.J.; Chen, X.L. Evaluations of the effective material properties of carbon nanotube-based composites using a nanoscale representative volume element. *Mech. Mater.* **2003**, *35*, 69–81. [CrossRef]
20. Fu, S.-Y.; Feng, X.-Q.; Lauke, B.; Mai, Y.-W. Effects of particle size, particle/matrix interface adhesion and particle loading on mechanical properties of particulate-polymer composites. *Compos. Part B Eng.* **2008**, *39*, 933–961. [CrossRef]
21. Guo, S.; Sivakumar, R.; Kitazawa, H.; Kagawa, Y. Electrical properties of silica-based nanocomposites with multiwall carbon nanotubes. *J. Am. Ceram. Soc.* **2007**, *90*, 1667–1670. [CrossRef]
22. Bao, H.-D.; Guo, Z.-X.; Yu, J. Effect of electrically inert particulate filler on electrical resistivity of polymer/multi-walled carbon nanotube composites. *Polymer* **2008**, *49*, 3826–3831. [CrossRef]
23. Park, S.-H.; Hwang, J.; Park, G.-S.; Ha, J.-H.; Zhang, M.; Kim, D.; Yun, D.-J.; Lee, S.; Lee, S.H. Modeling the electrical resistivity of polymer composites with segregated structures. *Nat. Commun.* **2019**, *10*, 2537. [CrossRef] [PubMed]

24. Volkov, A.N.; Shiga, T.; Nicholson, D.; Shiomi, J.; Zhigilei, L.V. Effect of bending buckling of carbon nanotubes on thermal conductivity of carbon nanotube materials. *J. Appl. Phys.* **2012**, *111*, 053501. [CrossRef]
25. Lekawa-Raus, A.; Patmore, J.; Kurzepa, L.; Bulmer, J.; Koziol, K. Electrical properties of carbon nanotube based fibers and their future use in electrical wiring. *Adv. Funct. Mater.* **2014**, *24*, 3661–3682. [CrossRef]
26. Li, C.; Thostenson, E.T.; Chou, T.-W. Dominant role of tunneling resistance in the electrical conductivity of carbon nanotube–based composites. *Appl. Phys. Lett.* **2007**, *91*, 223114. [CrossRef]
27. Palza, H.; Reznik, B.; Wilhelm, M.; Arias, O.; Vargas, A. Electrical, thermal, and mechanical characterization of poly (propylene)/carbon nanotube/clay hybrid composite materials. *Macromol. Mater. Eng.* **2012**, *297*, 474–480. [CrossRef]
28. Palza, H.; Garzón, C.; Arias, O. Modifying the electrical behaviour of polypropylene/carbon nanotube composites by adding a second nanoparticle and by annealing processes. *Express Polym. Lett.* **2012**, *6*, 639–646. [CrossRef]
29. Thongruang, W.; Spontak, R.J.; Balik, C.M. Bridged double percolation in conductive polymer composites: An electrical conductivity, morphology and mechanical property study. *Polymer* **2002**, *43*, 3717–3725. [CrossRef]

materials

MDPI

Article

Effect of Dispersion by Three-Roll Milling on Electrical Properties and Filler Length of Carbon Nanotube Composites

Ji-Hwan Ha [1], Sang-Eui Lee [2,*] and Sung-Hoon Park [1,*]

1 Department of Mechanical Engineering, Soongsil University, 369 Sangdo-ro, Dongjak-gu, Seoul 06978, Korea; jhwan618@gmail.com

2 Department of Mechanical Engineering, Inha University, Inha-ro 100, Michuhol-gu, Incheon 22212, Korea

* Correspondence: selee@inha.ac.kr (S.-E.L.); leopark@ssu.ac.kr (S.-H.P.); Tel.: +82-32-860-7377 (S.-E.L.); +82-2-828-7021 (S.-H.P.)

Received: 9 October 2019; Accepted: 12 November 2019; Published: 21 November 2019

Abstract: For practical use of carbon nanotube (CNT) composites, especially in electronic applications, uniform dispersion of a high concentration of CNTs in a polymer matrix is a critical challenge. Three-roll milling is one of most reliable dispersion techniques. We investigate the effect of three-roll milling time on CNT length and the electrical properties of a CNT/polydimethylsiloxane composite film with 10 wt% CNTs. During the milling process, the CNT length is decreased from 10 to 1–4 μm by mechanical shear forces. The electrical conductivity increases after 1.5 min of milling owing to dispersion of the CNTs but decreases with increasing milling time owing to the decrease in the CNT length. Considering the changes in the electrical conductivity of the CNT composite and CNT length, we determined how to optimize the three-roll milling time to obtain a suitable dispersion state.

Keywords: carbon nanotube; composite; three-roll milling; CNT dispersion; filler length variation

1. Introduction

Carbon-based nanomaterials are used to improve the electrical conductivity of composites owing to their superior electrical properties [1,2]. Many carbon-based nanomaterials are used as conductive fillers, for example, carbon black, carbon nanotubes (CNTs), and graphene. CNTs have been investigated for use in various applications owing to their superior electrical, thermal, and mechanical properties [3–5]. In addition, CNTs have a large aspect ratio and are chemically stable. Consequently, CNTs have been mixed with a polymer to fabricate conducting composites [6–8]. Tube-shaped conductive fillers can easily form connected networks in a polymer. The conductive pathway in the composite improve its electrical properties [9]. Furthermore, the aspect ratio of CNTs is related to their electrical properties. According to the literature, the electrical conductivity is enhanced by increasing the aspect ratio of the filler because the resulting electrical network includes many contact points between conductive fillers [10,11]. In addition, long CNTs can easily form a dense electrical network. Therefore, longer filler length could improve the electrical performance of composites.

To maximize the electrical and mechanical properties of composites, CNTs must be dispersed uniformly in the polymer matrix. However, nano-size carbon fillers such as CNTs, graphene, and carbon black often appear as large bundles or aggregates owing to the strong van der Waals forces, which produce a non-uniform dispersion [12–15]. Sonication is typically used as a traditional lab-scale method for dispersion of CNTs [16–20], in particular to disperse aggregated CNTs in solution. However, this method cannot be used when the filler concentration is high owing to the high viscosity of the CNT solution. In addition, over time, ultrasonically dispersed CNTs in solution re-aggregate owing to the low viscosity of the CNT solution. To address these limitations, many researchers have developed alternative dispersion methods. Three-roll milling is an effective method for dispersing a high filler content to improve many composites [4,21–26]. This method uses the mechanical shear forces by decreasing the gap between the rolls. The CNTs in the mixing paste are uniformly dispersed by passing through the microscale gap [22,23]. Furthermore, the number of three-roll mill passes (milling time) affects the CNT aspect ratio.

In this study, we fabricated CNT/polydimethylsiloxane (PDMS) composite films with a high CNT content (10 wt%) to obtain an electrically conductive composite. We used three-roll milling to obtain a uniform dispersion in the CNT composite. The effect of the milling time on the filler length was observed by scanning electric microscopy (SEM). In addition to the morphological analysis, Raman spectroscopy was used to determine the changes in the carbon structure of the CNTs. Furthermore, we measured the electrical properties associated with the length variation resulting from differences in the three-roll milling time.

2. Materials and Methods

Multi-walled CNTs (CM250, 15 nm in diameter and 10–15 μm in length, Hanwa Nanotech, Seoul, South Korea) were used as a conducting filler. PDMS was used as a polymer matrix (Dow Corning, Midland, MI, USA, Sylgard 184). A paste mixer (Daehwa, Seoul, Korea) and three-roll mill (Intech, Gyeonggi-do, Korea) were used to obtain a high-viscosity CNT paste. The CNT contents of the composite paste is 10 wt%. First, CNT and PDMS (elastomer/curing agent = 10:1) were premixed by a paste mixer at 500 rpm for 30 s and 1500 rpm for 60 s. Next, three-roll milling for various durations was conducted to disperse the CNTs. Seven composite pastes were prepared by milling for 0, 0.5, 1.5, 2, 3, and 6 min. The gaps between the rolls are 5 μm (front gap) and 12.5 μm (back gap) each. For electrical measurement of the composite, CNT composite films were fabricated by hot film pressing (Qmesys, Inc., Gyeonggi-do, Korea) at 150 °C for 40 min.

The CNT length after three-roll milling was observed using SEM (XL30, Phillips, North Billerica, MA, USA). Chloroform (solvent) was used to dissolve the PDMS in the CNT paste. Sonication was conducted at a frequency of 20 kHz for 20 min. After the sonication process (Branson 450, VWR, Radnor, PA, USA), the CNT solution was spin-coated on a Si wafer. The spin coating conditions were 500, 1500, and 500 rpm for 30, 60, and 30 s, respectively. The coated wafer was dried at room temperature overnight. The wafers that is coated the composite pastes were observed in 5 kV condition without conductive costing by SEM. For the measurement of CNTs length on the composite, we observed SEM images. Furthermore, we calculated the distributed length of CNTs according to three-roll milling time.

After CNT/PDMS film fabrication, UV etching was conducted for 300 s to ensure electrical contact by UV ozone cleaner (Jaesung engineering Co., Seoul, Korea). Silver paste (Protavic, Levallois-Perret, France) was used as an electrode; it was coated on the film and cured in an oven at 175 °C for 1 h. Next, a four-wire resistance method (Keithley 487 picoammeter and Keithley 2400 sourcemeter, Keithley, Cleveland, OH, USA) was used to measure the resistance of the composites.

3. Results and Discussion

The dispersion of CNTs in the polymer was performed by applying mechanical shear forces in three-roll milling. Figure 1 schematically illustrates the dispersion mechanism of the three-roll mill. Shear forces are applied by the microscale gaps between the rollers. CNTs are bundled by the van der Waals force before milling. However, the shear forces of the three-roll mill can produce a uniform dispersion of the fillers. Therefore, this method is typically used to disperse a carbon-based filler in a composite. CNT aggregation negatively affects the electrical, thermal, and mechanical properties of the composite. For example, agglomeration of CNTs in a composite can decrease the electrical conductivity because there are fewer contacts between CNTs in the electrical pathway of the composite. Therefore, a uniform dispersion is needed to improve the electrical properties. However, three-roll milling can have negative effects. Although it can disperse the conductive filler in the composite, increasing the roll milling time decreases the CNT aspect ratio. Specifically, when three-roll milling is performed for longer than 1.5 min, the CNTs become shorter as a result of high shear forces by the microgap in the three-roll mill. Therefore, the duration of dispersion by milling must be optimized to improve the electrical properties.

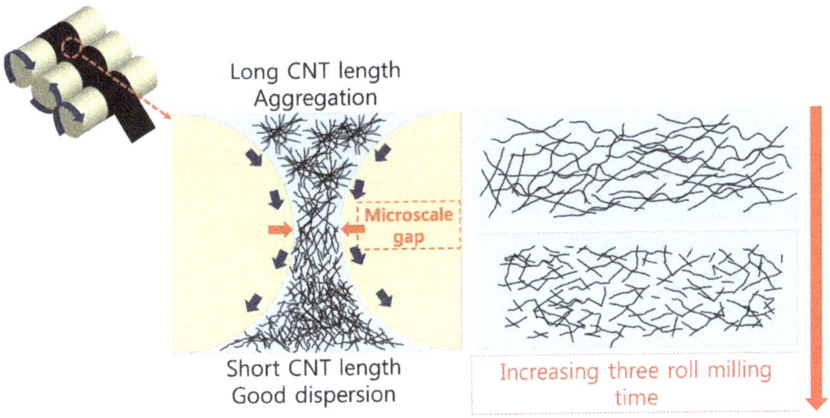

Figure 1. Mechanism of CNT (carbon nanotube) dispersion and aspect ratio changes by shear forces during three-roll milling. Filler length is reduced by increasing the three-roll milling time.

Figure 2 shows SEM images of CNTs in the CNT/PDMS paste. With no milling, the fillers are approximately 9–10 μm long. After 1.5 min of milling, the filler size decreases remarkably, as shown in Figure 2b. CNTs with lengths of 4–5 μm are dispersed on the wafer. When the dispersion time increases further to 3 min, the CNTs are even shorter (2.5–3 μm). Furthermore, the length of the filler is as short as 1–1.5 μm after 6 min of milling time. The length decreases with increasing dispersion time because three-roll milling applies pressure and shear forces to the CNTs.

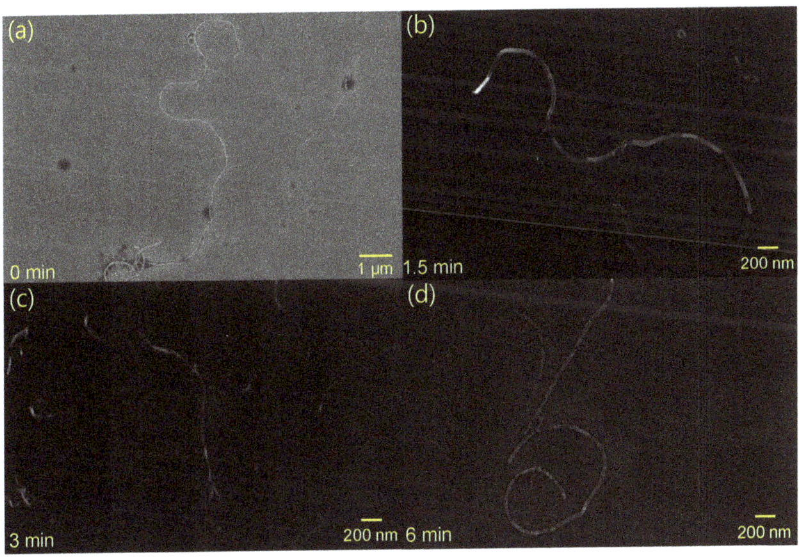

Figure 2. SEM images showing length of CNTs in CNT/PDMS paste after three-roll milling for (**a**) 0 min, (**b**) 1.5 min, (**c**) 3 min, and (**d**) 6 min.

To confirm the characteristics of the CNT composite, Raman spectra are obtained. Figure 3 shows the Raman spectra of pristine CNT and composites (6 min). The D band (representing defects and disorder) and G band (representing carbon structure) in the spectra are identical. In addition, the I_D/I_G ratios of pristine CNT and CNT/PDMS composite are 1.37 and 1.31, respectively. Therefore, the characteristics of the carbon atoms in the CNT filler are not changed (such as atomic destruction) by three-roll milling. Furthermore, the changes in filler length during three-roll milling do not affect the nature of the CNTs.

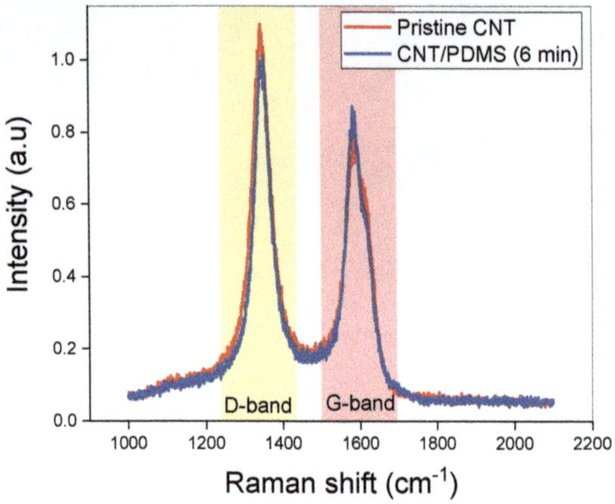

Figure 3. Raman spectra of pristine CNT and CNT composites.

Figure 4 shows the length distributions of CNTs milled for various durations. Without three-roll milling, the length of the CNTs is distributed almost entirely between 7 and 14 μm, and most of the CNTs are 10 μm long (this is the original length). However, the CNT size can be controlled by using three-roll milling. After milling for 1.5 min, the length decreases from 10 to 4 μm, and the length distribution ranges from 1 to 8 μm owing to the shear forces applied by the three-roll mill. The length is decreased to 0.5–3.5 μm by increasing the milling time to 3 min, and most of the CNTs are shorter (3 μm) than those obtained by 1.5 min of milling. The length of the CNTs milled for 6 min has the same range as that of the CNTs milled for 3 min (Figure 4d). However, most of the CNTs milled for 6 min have a length of 1 μm. Therefore, increasing the milling time decreases the CNT aspect ratio because more force is applied to the carbon fillers during a longer milling time.

Figure 5 shows SEM images that reveal the morphology (i.e., dispersion state) of CNT/PDMS films fabricated using three-roll milling for 1, 1.5, and 6 min. For 1 min of milling, the degree of dispersion is poor owing to the short milling time. For 1.5 min of milling, the CNTs are dispersed well owing to the moderate milling time. When the milling time is increased to 6 min, the composite morphology shows the most uniform dispersion. Therefore, a long three-roll milling time results in a uniform dispersion and short CNTs. Thus, when three-roll milling is used, the dispersion state and the CNT length have a trade-off relationship.

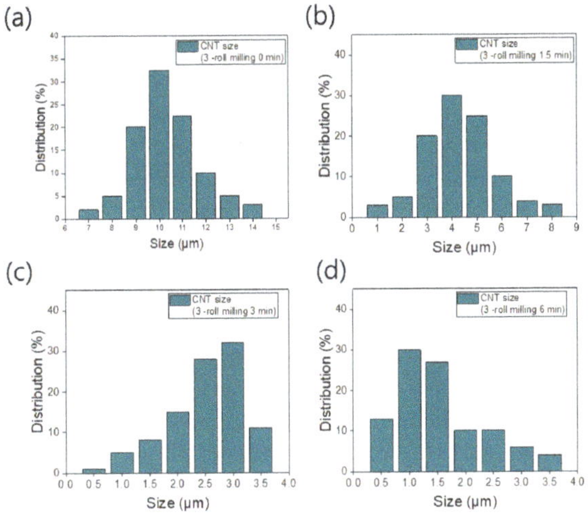

Figure 4. Length distributions of CNTs milled for (**a**) 0 min, (**b**) 1.5 min, (**c**) 3 min, and (**d**) 6 min.

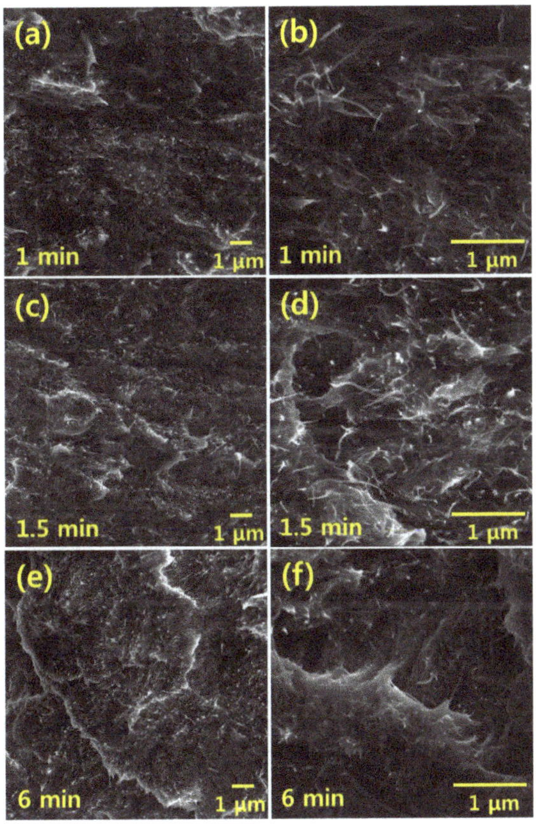

Figure 5. SEM images showing morphology of CNT/PDMS composites fabricated using three-roll milling for (**a**), (**b**) 1 min, (**c**), (**d**) 1.5 min, and (**e**), (**f**) 6 min.

We measured the electrical properties of composites obtained using various dispersion times. Figure 6a shows that the CNT length is decreased by increasing the milling time because of the mechanical shear forces. The error bar shows all CNT length distribution. As the milling time increases, the length of CNTs becomes more uniform. The length with no dispersion is distributed between 7 and 14 μm. Figure 6b shows the conductivity of CNT/PDMS films (CNT content: 10 wt%). The maximum conductivity of 372 S/m is measured after 1.5 min of milling time. Furthermore, this initial increase in the conductivity indicates that the composite contains many contacts owing to the large aspect ratio of the CNTs. For milling times longer than 1.5 min, the conductivity decreases because the aspect ratio of the fillers decreases. For milling durations of 2, 3, and 6 min, the electrical conductivity is 297, 280, and 248 S/m respectively. The number of contacts between fillers decreases because the short CNTs are not connected to each other. On the other hand, the non-uniform dispersion case (1 min milling) results in large variations in conductivity and relatively lower conductivity. With increasing milling time, the dispersion state becomes more stable, and CNT length decreases. The more stable dispersion state means homogenous dispersion of CNTs without aggregation of fillers. At this state, the aggregation of CNTs is not observed through SEM imaging. As a result, although the conductivity decreases, it varies less across the film. This result indicates that the aspect ratio of the filler and the dispersion state have a trade-off relationship during three-roll milling. Therefore, in terms of high conductivity with moderate error bar, 1.5 min milling time is optimal condition for the fabrication of the CNT/PDMS composites. In addition, the electrical conductivity and its variation can be easily tuned by controlling the three-roll milling time.

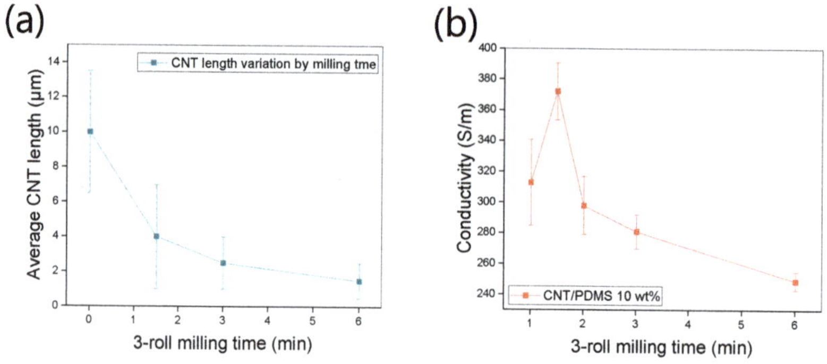

Figure 6. (**a**) Changes in CNT length with increasing three-roll milling time. (**b**) Electrical conductivity of CNT composite films obtained by milling for various times. The increase in electrical conductivity as the milling time increases from 1 to 1.5 min is due to the greater dispersion stability. After 1.5 min of milling time, the conductivity decreases because the decrease in CNT length reduces the number of CNT contacts.

4. Conclusions

We fabricated composite films with a high CNT content (10 wt%) to observe the effect of three-roll milling time on filler length and film electrical properties. The filler length and dispersion state are related to three-roll milling time. Dispersion times of 0, 1, 1.5, 3, and 6 min were applied. SEM images showed that as the milling time increased, the CNT length decreased, which was attributed to the application of mechanical forces. In addition, electrical conductivity was highest when the milling time was 1.5 min because a uniform dispersion of CNTs and a suitable filler length were obtained. However, the electrical conductivity decreased sharply when the milling time exceeded 1.5 min because the CNTs became shorter. There is optimal milling time for high conductivity and small conductivity deviation, which enable mass production of CNT composites.

Author Contributions: Conceptualization, S.-H.P.; methodology, J.-H.H. and S.-E.L.; formal analysis, S.-H.P. and S.-E.L.; resources, J.-H.H.; writing—original draft preparation, J.-H.H.; writing—review and editing, S.-E.L.; supervision, S.-H.P.; project administration, S.-H.P.; funding acquisition, S.-H.P.

Funding: This work was supported by Industrial Material Core Technology Development Program (Grant No 20004272, Development of Rapid Hand Cooling Technology for Injection Mold Using Nano Carbon based Surface Heating Element) funded by the Ministry of Trade, Industry & Energy.

Conflicts of Interest: The authors declare no conflict of interest.

References

1. Medalia, A.I. Electrical conduction in carbon black composites. *Rubber Chem. Technol.* **1986**, *59*, 432–454. [CrossRef]
2. Babaei, H.; Keblinski, P.; Khodadadi, J. Improvement in thermal conductivity of paraffin by adding high aspect-ratio carbon-based nano-fillers. *Phys. Lett. A* **2013**, *377*, 1358–1361. [CrossRef]
3. Chu, K.; Kim, D.; Sohn, Y.; Lee, S.; Moon, C.; Park, S. Electrical and thermal properties of carbon-nanotube composite for flexible electric heating-unit applications. *IEEE Electron Device Lett.* **2013**, *34*, 668–670. [CrossRef]
4. Hong, S.-K.; Kim, D.; Lee, S.; Kim, B.-W.; Theilmann, P.; Park, S.-H. Enhanced thermal and mechanical properties of carbon nanotube composites through the use of functionalized CNT-reactive polymer linkages and three-roll milling. *Compos. Part A* **2015**, *77*, 142–146. [CrossRef]
5. Alam, J.; Khan, A.; Alam, M.; Mohan, R. Electroactive shape memory property of a Cu-decorated CNT dispersed PLA/ESO nanocomposite. *Materials* **2015**, *8*, 6391–6400. [CrossRef]
6. Chu, K.; Lee, S.-C.; Lee, S.; Kim, D.; Moon, C.; Park, S.-H. Smart conducting polymer composites having zero temperature coefficient of resistance. *Nanoscale* **2015**, *7*, 471–478. [CrossRef]
7. Min, C.; Shen, X.; Shi, Z.; Chen, L.; Xu, Z. The electrical properties and conducting mechanisms of carbon nanotube/polymer nanocomposites: A review. *Polym. Plast. Technol. Eng.* **2010**, *49*, 1172–1181. [CrossRef]
8. Khare, R. Carbon nanotube based composites-a review. *J. Miner. Mater. Charact. Eng.* **2005**, *4*, 31–46. [CrossRef]
9. Pfeifer, S.; Park, S.-H.; Bandaru, P. Analysis of electrical percolation thresholds in carbon nanotube networks using the Weibull probability distribution. *J. Appl. Phys.* **2010**, *108*, 024305. [CrossRef]
10. Menon, M.; Andriotis, A.N.; Srivastava, D.; Ponomareva, I.; Chernozatonskii, L.A. Carbon nanotube "T junctions": Formation pathways and conductivity. *Phys. Rev. Lett.* **2003**, *91*, 145501. [CrossRef]
11. Li, J.; Ma, P.C.; Chow, W.S.; To, C.K.; Tang, B.Z.; Kim, J.K. Correlations between percolation threshold, dispersion state, and aspect ratio of carbon nanotubes. *Adv. Funct. Mater.* **2007**, *17*, 3207–3215. [CrossRef]
12. Zhu, L.; Xu, J.; Xiu, Y.; Sun, Y.; Hess, D.W.; Wong, C. Growth and electrical characterization of high-aspect-ratio carbon nanotube arrays. *Carbon* **2006**, *44*, 253–258. [CrossRef]
13. Huang, Y.Y.; Terentjev, E.M. Dispersion of carbon nanotubes: Mixing, sonication, stabilization, and composite properties. *Polymers* **2012**, *4*, 275–295. [CrossRef]
14. Lee, S.-E.; Park, S.-H. Enhanced dispersion and material properties of multi-walled carbon nanotube composites through turbulent Taylor-Couette flow. *Compos. Part A* **2017**, *95*, 118–124. [CrossRef]
15. Pourasghar, A.; Yas, M.; Kamarian, S. Local aggregation effect of CNT on the vibrational behavior of four-parameter continuous grading nanotube-reinforced cylindrical panels. *Polym. Compos.* **2013**, *34*, 707–721. [CrossRef]
16. Koh, B.; Cheng, W. Mechanisms of carbon nanotube aggregation and the reversion of carbon nanotube aggregates in aqueous medium. *Langmuir* **2014**, *30*, 10899–10909. [CrossRef]
17. Badard, M.; Combessis, A.; Allais, A.; Flandin, L. CNT aggregation mechanisms probed by electrical and dielectric measurements. *J. Mater. Chem. C* **2015**, *3*, 5769–5774.
18. Kim, J.A.; Seong, D.G.; Kang, T.J.; Youn, J.R. Effects of surface modification on rheological and mechanical properties of CNT/epoxy composites. *Carbon* **2006**, *44*, 1898–1905. [CrossRef]
19. Yu, S.; Zheng, W.; Yu, W.; Zhang, Y.; Jiang, Q.; Zhao, Z. Formation mechanism of β-phase in PVDF/CNT composite prepared by the sonication method. *Macromolecules* **2009**, *42*, 8870–8874. [CrossRef]
20. Ma, P.-C.; Siddiqui, N.A.; Marom, G.; Kim, J.-K. Dispersion and functionalization of carbon nanotubes for polymer-based nanocomposites: A review. *Compos. Part A* **2010**, *41*, 1345–1367. [CrossRef]

21. Liu, C.-X.; Choi, J.-W. Improved dispersion of carbon nanotubes in polymers at high concentrations. *Nanomaterials* **2012**, *2*, 329–347. [CrossRef]
22. Arrigo, R.; Teresi, R.; Gambarotti, C.; Parisi, F.; Lazzara, G.; Dintcheva, N. Sonication-induced modification of carbon nanotubes: Effect on the rheological and thermo-oxidative behaviour of polymer-based nanocomposites. *Materials* **2018**, *11*, 383. [CrossRef]
23. Park, S.-H.; Lee, S.; Moreira, D.; Bandaru, P.R.; Han, I.; Yun, D.-J. Bioinspired superhydrophobic surfaces, fabricated through simple and scalable roll-to-roll processing. *Sci. Rep.* **2015**, *5*, 15430. [CrossRef] [PubMed]
24. Chu, K.; Park, S.-H. Electrical heating behavior of flexible carbon nanotube composites with different aspect ratios. *J. Ind. Eng. Chem.* **2016**, *35*, 195–198. [CrossRef]
25. Theilmann, P.; Yun, D.-J.; Asbeck, P.; Park, S.-H. Superior electromagnetic interference shielding and dielectric properties of carbon nanotube composites through the use of high aspect ratio CNTs and three-roll milling. *Org. Electron.* **2013**, *14*, 1531–1537. [CrossRef]
26. Park, S.-H.; Hwang, J.; Park, G.-S.; Ha, J.-H.; Zhang, M.; Kim, D.; Yun, D.-J.; Lee, S.; Lee, S.H. Modeling the electrical resistivity of polymer composites with segregated structures. *Nat. Commun.* **2019**, *10*, 2537. [CrossRef]

materials

MDPI

Article

Strain-Sensing Properties of Multi-Walled Carbon Nanotube/Polydimethylsiloxane Composites with Different Aspect Ratio and Filler Contents

Oh-Nyoung Hur, Ji-Hwan Ha and Sung-Hoon Park *

Department of Mechanical Engineering, Soongsil University, 369 Sangdo-ro, Dongjak-Gu, Seoul 06978, Korea; ohnyung324@soongsil.ac.kr (O.-N.H.); jhwan618@ssu.ac.kr (J.-H.H.)
* Correspondence: leopark@ssu.ac.kr; Tel.: +82-2-828-7021

Received: 30 April 2020; Accepted: 19 May 2020; Published: 26 May 2020

Abstract: For filler composite systems used in strain sensor applications, piezoresistive effect, strain hysteresis, and repeatability are critical factors, which have to be clearly evaluated and understood. To investigate the effects of the aspect ratio and content of a multi-walled carbon nanotube (MWCNT) on the strain sensor properties of the composite, MWCNT/Polydimethylsiloxane (PDMS) composites with varying filler contents and aspect ratios were fabricated. In order to uniformly disperse MWCNTs on the polymer matrix, we used a three-roll milling method to generate high shear force for de-bundling MWCNTs. Mechanical and electrical properties of the MWCNT composites were evaluated for each case. In addition, through the cyclic stretching test, we optimized the strain-sensing properties of the MWCNT composites by considering their piezoresistive effects and strain hysteresis.

Keywords: carbon nanotube; aspect ratio; polymer composite; strain sensor; hysteresis

1. Introduction

Currently, various studies on pressure sensors and strain sensors are in progress [1–4]. Most commercially available strain sensors are metal-based and have a disadvantage in that their strain-sensing range is very limited. To improve the strain-sensing range, many studies have been aimed at developing a strain sensor with a wide strain-sensing range by adding various carbon-based fillers such as carbon black, carbon fiber, MWCNT, and graphene to the polymer [5–11]. MWCNT, one of the diverse fillers, is frequently used as a conductive filler because of its large aspect ratio. Due to the large aspect ratio of MWCNT, sufficient electrical paths are generated inside the composites even with the use of a small amount of MWCNT [12–14].

The piezoresistive effect is a key contributor to the sensing mechanism of the strain sensor. Piezoresistive effect describes a change in electrical resistance as a result of mechanical deformation [15]. Previous studies have been conducted to explore the factors influencing piezoresistive effect. From the results of those studies, piezoresistive effect is influenced by the dimensions and contents of the filler and the structure of the composite [3,6,16–18]. Some studies observed that electrical resistance did not return to the initial state even after the removal of the tension causing it from the composites [10,19]. This phenomenon is called hysteresis. In detail, hysteresis is a phenomenon in which a physical quantity is not determined only by the physical conditions at that time but also by the change process of the state that the substance has been in before. Former studies have shown that hysteresis is also affected by the dimension and content of fillers and the morphology of composites [9,20].

As seen above, the morphology of the composite is an important factor in determining both the piezoresistive effect and hysteresis; therefore, it is important to uniformly disperse conductive fillers on the composite. Ultra-sonication is a general method for dispersing MWCNTs on composites [16,21].

The method shows excellent dispersing effectiveness but poses a challenge with the dispersion of fillers on high viscosity composites. A solvent may be used to disperse fillers on high viscosity composites, but it causes inconvenience in terms of solvent removal during composite fabrication. Another method for conductive filler dispersion in high-viscosity composites is the three-roll mill method. This is a method of filler dispersion on composites that uses the shear force generated by three-rolls [22].

In this study, we fabricated 0.07, 0.5, 1, 3, and 5 wt% MWCNT polymer composites using short MWCNTs and 0.4, 0.5, 1, 3, and 5 wt% composites using long MWCNTs to determine the effects of the aspect ratios and content of MWCNTs on the mechanical and electrical properties of composites during strain sensing. We used the three-roll milling method to attain even dispersion at high viscosity. Figure 1 schematically presents the mechanism of the three-roll milling method and the morphology of two types of MWCNT composites. In order to calculate the aspect ratios of the two different conductive fillers, their lengths were measured by scanning electric microscopy (SEM). In addition, the internal morphologies of the composites were also observed via SEM. Mechanical and electrical properties of MWCNT composites were measured by tensile and cyclic experiments, respectively.

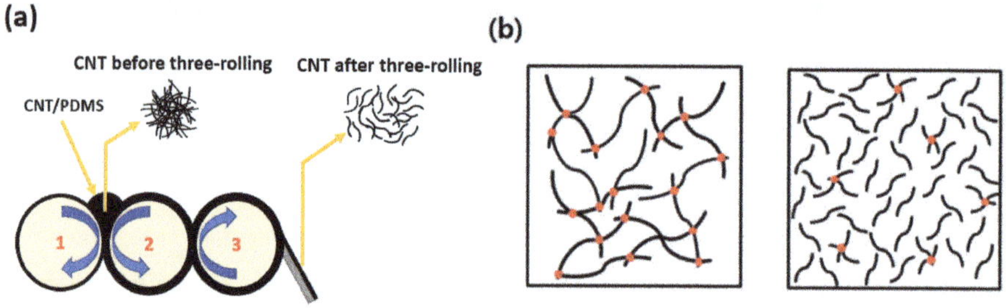

Figure 1. A scheme of (**a**) a carbon nanotube (CNT) dispersion mechanism during three-roll milling. (**b**) Morphology of long CNT composite (left) and short CNT composite (right) with the same filler content. Red dots indicate the electrical contact points.

2. Materials and Methods

2.1. Materials

Two types of MWCNTs were used as conductive fillers. The long MWCNT was purchased from JEIO (Incheon, Korea) and had a mean diameter of 5 nm, bundle length of 50–150 μm, and purity of >97.5 wt%. The short MWCNT was purchased from KB-Element Co., Ltd (Gyeonggi-do, Korea) and had a mean diameter of 5 nm, a length of 10–20 μm, and purity of >98 wt%. PDMS (Dow Corning, Sylgard 184, Midland, MI, USA) was adopted as the base polymer.

2.2. Fabrication of MWCNT/PDMS Composite

The paste was fabricated using a paste mixer (Daehwa, Seoul, Korea) and three-roll milling (Intech, Gyeonggi-do, Korea) for even dispersion of MWCNT on PDMS. The specific process was as follows: PDMS was first prepared with premixing elastomer(A) and curing agent(B) (mass ratio A:B=10:1), and then MWCNT was added to the PDMS. A paste mixer was then used to blend the paste at 500 revolutions per minute (rpm) for 30 s. Next, the paste was continuously mixed at 1500 rpm for 60 s. The MWCNTs were then uniformly dispersed on the composites by three-roll milling for 5 min. To measure the mechanical and electrical properties of the MWCNT/PDMS composites, a thin film was then made. The paste was further pressed and cured by a hot-press heating plate (Qmesys Inc., Gyeonggi-do, Korea), operated at 150 °C and 15 MPa for 1 h to obtain a thin and flat film with a thickness of 1 mm.

2.3. Characterization

To measure the length of the MWCNT, small quantities of two kinds of MWCNTs were separately dispersed in chloroform (Daejung, Seoul, Korea) by ultra-sonication at a pulse of 20 kHz for 25 min. Next, the suspension was spin-coated on a silicon wafer in a three-step process. The first step was conducted at 500 rpm for 5 s. The second step was conducted at 2000 rpm for 30 s, while the third step was conducted at 500 rpm for 5 s. The spin-coated wafers were then dried at an ambient temperature for a day. The dried spin-coated wafers were then observed by SEM (Gemini SEM 300, ZEISS Inc., Land Baden-Württemberg, Germany). The equipment was operated at an accelerating voltage of 5 kV. To understand the morphology of MWCNT/PDMS composites, the composites were fractured in liquid nitrogen. The cross-section of the fractured composites was coated with Pt and then observed using SEM at a 5 kV accelerating voltage.

The mechanical properties of the composites were measured using a universal testing machine (UTM) (DRTECH Inc., Gyeonggi-do, Korea). Mechanical test samples of 1 and 5 wt% MWCNT/PDMS composites with a thickness of 1 mm were made according to ASTMD 638-5. All mechanical tests were performed at a cross-head speed of 50 mm/min and with a load-cell of 50 kgf (490 N), and five specimens were tested.

To measure the electrical properties and percolation thresholds of the composites, various content composites were made. For each content composite, five samples were made, and the average values of the electrical properties and percolation thresholds were obtained. The sample size was $50 \times 5 \times 1$ mm^3. The samples were ultraviolet (UV) etched for 300 s by a UV ozone room (JSE Co., Seoul, Korea) to enhance the electrical contact between the samples and the silver paste (Protavic, Levallois-Perret, France). The silver paste was then covered on both ends of the surface of the samples as electrodes. After this process, the samples were cured in a stove at 120 °C for 1 h. After that, their resistance was determined using a two-wire method (Keithley DMM 7510 multi-meter, Keithley, Cleveland, OH, USA).

The hysteresis property tests were conducted using a 3-dimensional stretching machine (NAMIL optical instruments Co., Incheon, Korea) and a two-wire method. The samples were stretched from steady state to 30% strain and then released back to steady state at the same speed; the changes in their resistances were measure simultaneously. The released state was maintained for 1 min. The entire process was repeated over 30 cycles. The piezoresistive property tests were conducted immediately after the hysteresis property tests. The samples were stretched from steady state to 30% strain and then released to steady state continuously over 20 cycles without a rest time. During the stretching and releasing cycles, the resistances were measured in the same way as the resistance measurements conducted for the hysteresis tests.

3. Results and Discussion

3.1. Morphological Analysis

Figure 2 shows the SEM images of the two kinds of MWCNTs. In order to obtain SEM images of a single MWCNT, we physically separated them from the bundles by ultra-sonication. We took enough SEM images to determine the average length of each MWCNT. From Figure 2a,b, we calculated the average length of the longer MWCNT, which was found to be 12 μm with an aspect ratio of 2400. The MWCNT was named the high aspect ratio MWCNT (HCNT). We measured the mean length of the shorter MWCNT in Figure 2c,d and determined it to be 4.5 μm with an aspect ratio of 900. This was named the low aspect ratio MWCNT (LCNT). Furthermore, we found that HCNT is more twisted than LCNT.

We obtained the SEM images to confirm that the MWCNTs were uniformly dispersed on the composites by three-roll milling and how the internal morphology of the composite was constructed according to the aspect ratios and content of MWCNTs. Figure 3 shows SEM images of morphologies of the 1 and 5 wt% LCNT/PDMS and HCNT/PDMS composites, respectively. In Figure 3a,b, we observe that both LCNT and HCNT were uniformly dispersed on the composites.

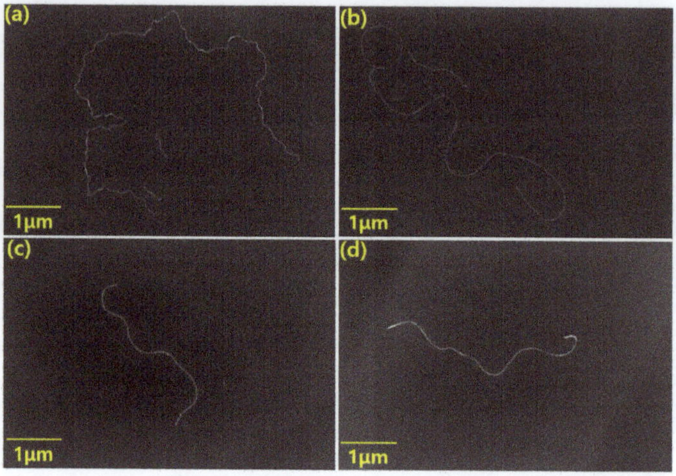

Figure 2. SEM images of MWCNTs: (**a**,**b**) in the case of high aspect ratio MWCNT (HCNT) with an average length of 12 μm and (**c**,**d**) in the case of low aspect ratio MWCNT (LCNT) with an average length of 4.5 μm.

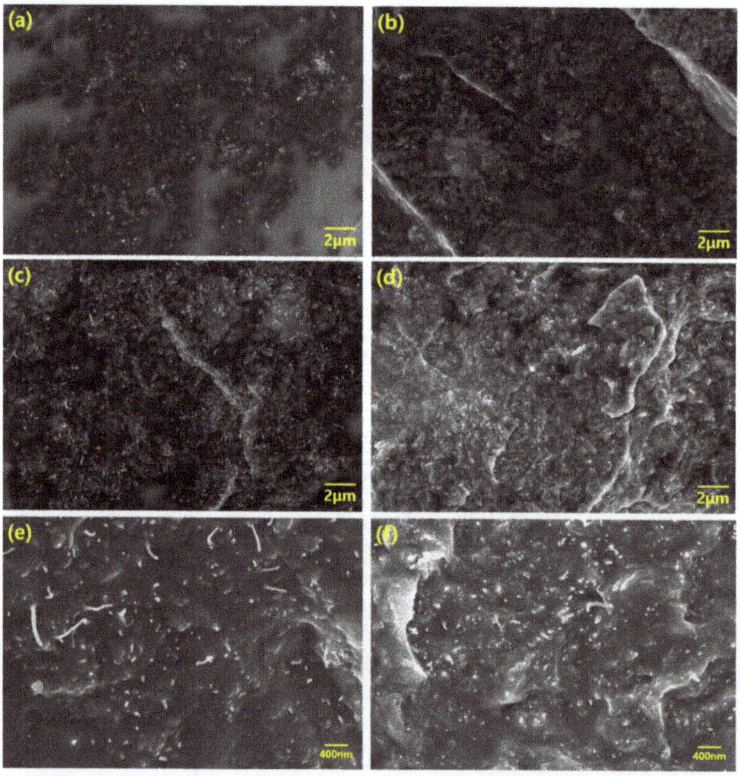

Figure 3. SEM images of MWCNT/PDMS composites: (**a**) 1 wt% LCNT/PDMS, (**b**) 1 wt% HCNT/PDMS, (**c**) 5 wt% LCNT/PDMS, (**d**) 5 wt% HCNT/PDMS with low resolution, and (**e**) 5 wt% LCNT/PDMS, (**f**) 5 wt% HCNT/PDMS with high resolution.

We discerned that high contents of both LCNT/PDMS and HCNT/PDMS are also evenly dispersed in Figure 3c,d. From Figure 3e,f, we observe that the fillers were more densely packed into the HCNT/PDMS composite than into the LCNT/PDMS composite at the same PDMS content. This difference in the density of filler packing is attributed to the longer length of HCNT relative to that of LCNT, which makes HCNT observation more likely.

3.2. Mechanical Properties

Tensile tests were conducted to find out the mechanical properties according to the aspect ratio and content of MWCNTs. Figure 4 shows the mechanical properties for each composite. We obtained Young's modulus(Elastic modulus) using Hooke's law, as given by

$$\sigma = \varepsilon E \tag{1}$$

where σ is the stress, ε is the strain, and E is Young's modulus. In Figure 4a, we observed that Young's modulus increased as the filler content and aspect ratio increased. This finding means that composites are more mechanically reinforced at higher aspect ratios and higher MWCNT contents. The reason for this is that the MWCNTs tend to bridge the polymer cracks. In this respect, HCNT is more advantageous than LCNT due to its greater length. Furthermore, load transfer from polymer to MWCNT occurs, mechanically reinforcing the composites. The longer length of HCNT also makes it more advantageous for load transfer compared to LCNT. The bridge effect and the load transfer of CNT were confirmed in previous papers [23–25]. Moreover, assuming that the filler is uniformly dispersed, increased filler content naturally increases the mechanical reinforcement of composites because of the increased load transfer and polymer bridge effects mentioned above.

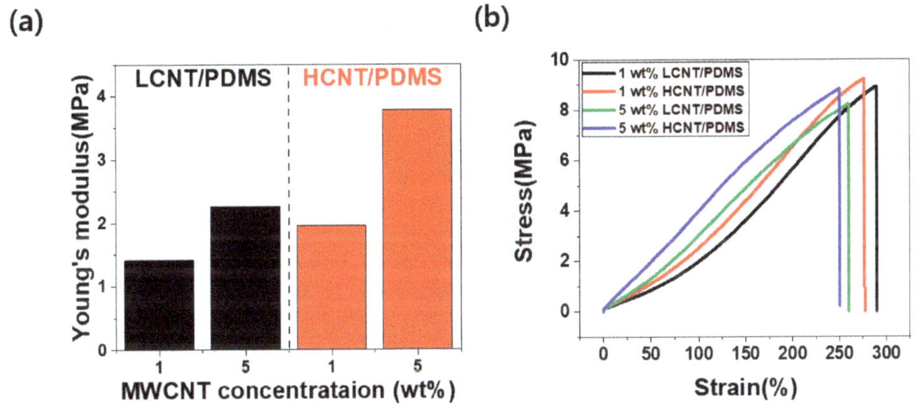

Figure 4. Mechanical properties of MWCNT/PDMS composites: (**a**) Young's modulus of LCNT/PDMS composites (black bar) and HCNT/PDMS composites (red bar), (**b**) strain–stress curve of the 1 and 5 wt% LCNT/PDMS and HCNT/PDMS composites.

However, Figure 4b shows that tensile elongation at break (ε_{break}) decreased with a higher aspect ratio of MWCNTs and larger content of MWCNTs. In the 1 wt% composites, ε_{break} decreased from 290.22% (LCNT/PDMS) to 277.48% (HCNT/PDMS) with an increasing aspect ratio of MWCNTs. In the 5 wt% composites, ε_{break} decreased from 260.65% (LCNT/PDMS) to 251.05% (HCNT/PDMS) with an increasing aspect ratio of MWCNTs. Furthermore, as the filler content increased, the tensile stress decreased slightly. These phenomena were caused by a shortage of interfacial interplay between the MWCNTs and PDMS. Namely, MWCNTs are likely to exist in the bundle state or in aggregated form at high content, hence resulting in the poor interaction between MWCNT and PDMS. Therefore, the tensile stress decreased slightly, as shown in Figure 4b. As the filler content increased, the reduction of tensile stress and the decrease in the interaction between filler and polymer were also observed in previous studies [26].

3.3. Electrical Conductivity and Percolation Threshold

The electrical properties of the composite are mainly determined by the aspect ratio and content of the filler. Figure 5 shows the variation in the electrical conductivity of LCNT/PDMS and HCNT/PDMS with content. The electrical conductivity ($\sigma_{electrical}$) is calculated by Equation (2).

$$\sigma_{electrical} = \frac{\ell}{SR} \quad (2)$$

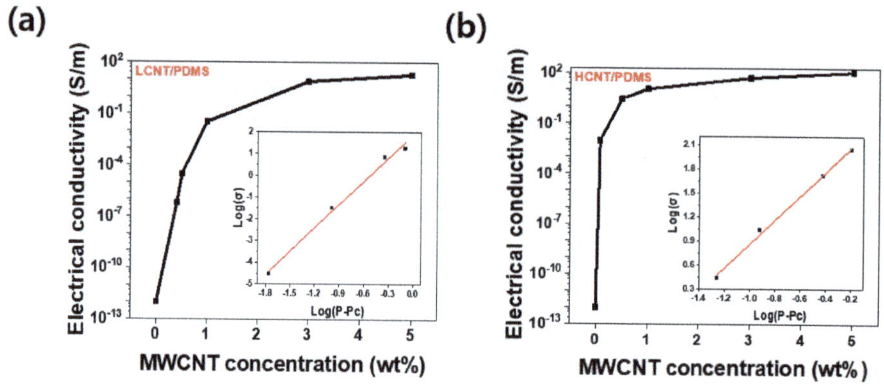

Figure 5. The electrical conductivity of MWCNT/PDMS composites as a function of MWCNT wt% and electrical percolation threshold of MWCNT/PDMS composites. Inset: log-log plot of the conductivity of composites according to the relation ($P-P_c$): (**a**) LCNT/PDMS composites case and (**b**) HCNT/PDMS composites case.

In Equation (2), ℓ is the length between the electrodes, S is the cross-section of the sample, and R is the electrical resistance of the sample. We observed that at the same content, HCNT/PDMS has a higher electrical conductivity than LCNT/PDMS. This is because HCNT has a greater length, making it advantageous for the formation of conductive paths inside the composites. Previous papers also showed that fillers with higher aspect ratios have larger conductivity at the same content [14,16].

Furthermore, sharp increases in electrical conductivity were observed at 0.4 wt% for LCNT/PDMS and at 0.07 wt% for HCNT/PDMS. This phenomenon was explained by the percolation theory [27,28]. The power law for $P > P_c$ is shown in Equation (3) below, where σ is the electrical conductivity of the composites, σ_z is the reference electrical conductivity, P is filler content, P_c is percolation threshold, and t is a critical exponent.

$$\sigma = \sigma_z(P - P_c)^t \quad (3)$$

From Equation (3), considering σ_z as a constant value, we obtained that Pc = 0.4 wt% and t = 3.58 for LCNT/PDMS, whereas P_c = 0.07 wt% and t = 1.49 for HCNT/PDMS. Calculated values of P_c and t for the LCNT/PDMS and HCNT/PDMS composites followed the trends of the same values observed in previous studies [16,26,29–31]. We observed that HCNT/PDMS had a lower percolation threshold than LCNT/PDMS. HCNT has the advantage of the formation of conductive paths as it is longer than LCNT. The advantage comes in that even at lower filler contents, conductive paths are effectively formed; hence, HCNT/PDMS has a lower percolation threshold than LCNT/PDMS. Furthermore, electrical conductivity tended to be more saturated as the filler content increased. This is because enough conductive paths were already formed on the composites above a certain filler content, so that the addition of more filler results in only weak improvements in the electrical conductivity.

3.4. Piezoresistive Properties in Cyclic Loading

Figure 6 shows the $\Delta R/R_0$ (relative change in resistance) of the sample under tensile deformation (where R is the real-time resistance of the samples, R_0 is the initial resistance of the samples, and ΔR is $R-R_0$). In Figure 6, when the 1 wt% and 5 wt% LCNT/PDMS and HCNT/PDMS composite samples were stretched to 30% strain, the relative change in resistance decreased from 123% to 70%, from 39% to 13%, from 42% to 34%, and from 29% to 11%, respectively, as the number of cycles increased, and the relative change in resistance was finally saturated after more than 20 discontinuous cycles. The hysteresis values of 1 wt% and 5 wt% of LCNT/PDMS and HCNT/PDMS also reduced from 12% to 5%, from 23% to 2%, from 11% to 6%, and from 14% to 8%, respectively, and they were then eventually saturated as well.

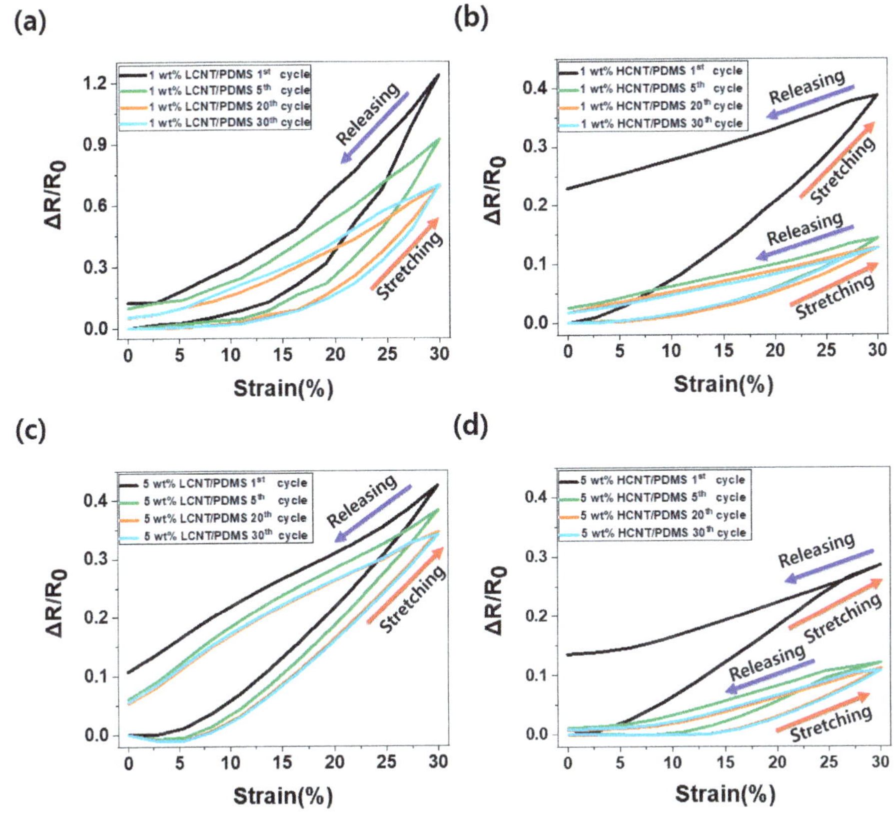

Figure 6. Normalized change in resistance ($\Delta R/R_0$) versus strain for MWCNTs/PDMS composite under discontinuous cyclic tensile deformation: (**a**) The case of 1 wt% LCNT/PDMS composite, (**b**) the case of 1 wt% HCNT/PDMS composite, (**c**) the case of 5 wt% LCNT/PDMS composite, and (**d**) the case of 5 wt% HCNT/PDMS composite.

Here, we observed the decrease of relative change in resistance following the repeated cycles. Moreover, the values of relative change in resistance and hysteresis were eventually saturated when the number of cycles exceeded 20. This is because the morphology of MWCNTs inside the composites is rearranged with repeated constant stretching and releasing. The relative change in resistance and hysteresis value changed depending on the morphology of the filler, and the change in the morphology of the fillers was saturated under repeated constant strain and releasing cycles. Therefore, relative change in resistance and the hysteresis value were stabilized depending on the stabilization of the filler morphology.

Comparing the relative change in resistance of LCNT/PDMS and HCNT/PDMS, we observed that LCNT/PDMS showed a better piezoresistive effect and stable hysteresis properties after the 1st cycle. The reason for the better piezoresistive properties of LCNT/PDMS is that LCNT is shorter than HCNT, so that the connections between LCNTs are easily broken during tension. Furthermore, the wavy HCNT expanded during tension, thus exhibiting an advantage by maintaining contacts between the HCNTs. According to previous literature, the expansion of the wavy HCNT negatively affects its piezoresistive effect [32,33]. Therefore, a better piezoresistive effect is observed in LCNT/PDMS. In terms of hysteresis properties, LCNT/PDMS shows smaller hysteresis properties than HCNT/PDMS because LCNT has a lower tendency for buckling than HCNT upon release from tensile deformation. Theoretical analyses presented in previous papers confirm this phenomenon [20]. In other words, the electrical paths inside the LCNT/PDMS are less restructured than the electrical paths inside the HCNT/PDMS when released after stretching.

Figure 7 shows the relative changes in resistance according to the aspect ratio and filler contents for 20 cycles of continuous stretching and releasing. The part where the relative change in resistance increases is the stretching state (up to ~30% strain), and the part where the relative change in resistance decreases is the releasing state. We observed that the relative change in resistance increased as the filler content decreased. This is because the contacts between the MWCNTs inside the composites are more likely to break during stretching at low filler content.

When MWCNTs of low filler content composites are rearranged during tensile stretching, new contacts are less likely to occur than in composites with higher filler contents. Namely, the relative change in resistance is large in the composites with low filler content because the morphology of the MWCNTs changes due to the tension, and there is more contact loss than formation of new contacts. Furthermore, according to the antecedent papers, contact loss results in increasing tunneling resistance and has a positive effect on the piezoresistive effect [34,35].

As the filler content increased, the relative change in resistance in the LCNT/PDMS composite decreased from 69% to 29%, a decrease of approximately 58%, whereas the relative change in resistance in the HCNT/PDMS composite decreased from 13% to 11%, a change of approximately 2%. The difference in the relative change in resistance between the 1 wt% and the 5 wt% LCNT/PDMS is greater than that in HCNT/PDMS. As we saw above, HCNT/PDMS has a low percolation threshold (at 0.07 wt%), so sufficient conductive paths exist even at 1 wt% composite, whereas LCNT/PDMS has a large percolation threshold (at 0.4 wt%), and it results in the insufficiency of the conductive paths at the same wt% composite. Therefore, the increased content, from 1 to 5 wt%, would have a great effect on the conductive paths in the LCNT/PDMS. On the other hand, the conductive paths in the HCNT/PDMS were not significantly affected by the increment. Moreover, we observed stable and constant relative change in resistance over 20 cycles of continuous stretching and releasing.

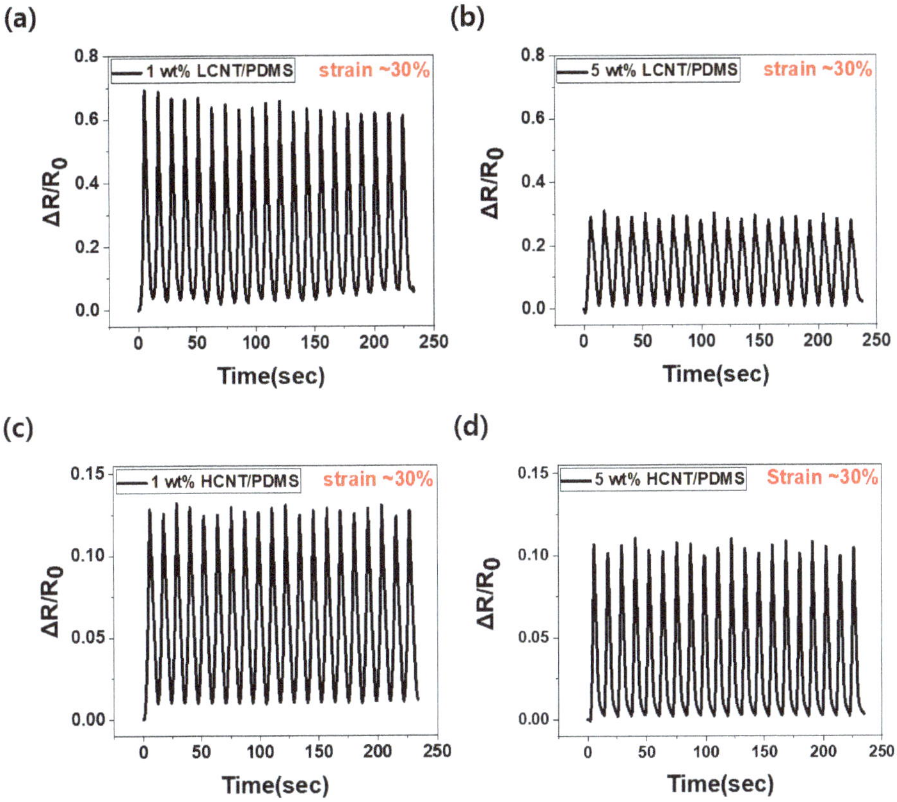

Figure 7. Normalized change in resistance ($\Delta R/R_0$) under 20 cycles of continuous tensile deformations to 30% strain: (**a**) Case of 1 wt% LCNT/PDMS, (**b**) case of 5 wt% LCNT/PDMS, (**c**) case of 1 wt% HCNT/PDMS, and (**d**) case of 5 wt% HCNT/PDMS. (For all samples, 30 pre-discontinuous cycles were executed before the continuous cyclic test).

4. Conclusions

To investigate the strain-sensing properties according to the aspect ratio and content of MWCNTs within composites, 0.4, 0.5, 1, 3, and 5 wt% LCNT/PDMS and 0.07, 0.5, 1, 3, and 5 wt% HCNT/PDMS composites were fabricated. From the tensile tests, we confirmed that using low filler content with LCNT resulted in a lower Young's modulus and a larger strain range than using high filler content with HCNT due to the bridge effect of the filler and loading transfer. In addition, by measuring the electrical conductivity, we proved that LCNT/PDMS had a lower electrical conductivity and a larger percolation threshold than HCNT/PDMS because of the poor ability of LCNT to create electrical paths in the composites. However, through the discontinuous cycle tests, we confirmed that LCNT/PDMS exhibited a more sensitive piezoresistive effect and stable hysteresis properties than HCNT/PDMS because LCNT has a short length while HCNT is characterized by waviness. Furthermore, piezoresistive effect and hysteresis value were saturated according to the repeated constant strain and releasing. Therefore, for the stable use of the strain sensor, pre-stretching is required in a larger range of strains than the strain section we will apply. Finally, through continuous cycle tests, we found that low filler content resulted in a stronger piezoresistive effect than high filler content since there is greater total conductive path loss at lower filler content than at higher filler content composite during stretching.

In conclusion, composites with lower aspect ratio and smaller filler content are great condition in terms of piezoresistive effect and strain hysteresis.

Author Contributions: Conceptualization, S.-H.P.; methodology, O.-N.H. and J.-H.H.; formal analysis, S.-H.P.; resources, O.-N.H. and J.-H.H.; writing—original draft preparation, O.-N.H.; writing—review and editing, S.-H.P.; supervision, S.-H.P.; project administration, S.-H.P.; funding acquisition, S.-H.P. All authors have read and agreed to the published version of the manuscript.

Funding: This work was supported by the National Research Foundation of Korea (NRF) grant funded by the Korean government (MSIT). (No. 2020R1A2C1013489)

Conflicts of Interest: The authors declare no conflict of interest.

References

1. Atalay, O. Textile-based, interdigital, capacitive, soft-strain sensor for wearable applications. *Materials* **2018**, *11*, 768. [CrossRef] [PubMed]
2. Kim, S.Y.; Park, S.; Park, H.W.; Park, D.H.; Jeong, Y.; Kim, D.H. Highly sensitive and multimodal all-carbon skin sensors capable of simultaneously detecting tactile and biological stimuli. *Adv. Mater.* **2015**, *27*, 4178–4185. [CrossRef] [PubMed]
3. Li, Q.; Li, J.; Tran, D.; Luo, C.; Gao, Y.; Yu, C.; Xuan, F. Engineering of carbon nanotube/polydimethylsiloxane nanocomposites with enhanced sensitivity for wearable motion sensors. *J. Mater. Chem.* **2017**, *5*, 11092–11099. [CrossRef]
4. Lou, C.; Wang, S.; Liang, T.; Pang, C.; Huang, L.; Run, M.; Liu, X. A graphene-based flexible pressure sensor with applications to plantar pressure measurement and gait analysis. *Materials* **2017**, *10*, 1068. [CrossRef] [PubMed]
5. Lee, J.-B.; Khang, D.-Y. Electrical and mechanical characterization of stretchable multi-walled carbon nanotubes/polydimethylsiloxane elastomeric composite conductors. *Compos. Sci. Technol.* **2012**, *72*, 1257–1263. [CrossRef]
6. Liu, H.; Gao, J.; Huang, W.; Dai, K.; Zheng, G.; Liu, C.; Shen, C.; Yan, X.; Guo, J.; Guo, Z. Electrically conductive strain sensing polyurethane nanocomposites with synergistic carbon nanotubes and graphene bifillers. *Nanoscale* **2016**, *8*, 12977–12989. [CrossRef]
7. Zheng, Y.; Li, Y.; Dai, K.; Wang, Y.; Zheng, G.; Liu, C.; Shen, C. A highly stretchable and stable strain sensor based on hybrid carbon nanofillers/polydimethylsiloxane conductive composites for large human motions monitoring. *Compos. Sci. Technol.* **2018**, *156*, 276–286. [CrossRef]
8. Tsouti, V.; Mitrakos, V.; Broutas, P.; Chatzandroulis, S. Modeling and development of a flexible carbon black-based capacitive strain sensor. *IEEE Sens. J.* **2016**, *16*, 3059–3067. [CrossRef]
9. Montazerian, H.; Dalili, A.; Milani, A.; Hoorfar, M. Piezoresistive sensing in chopped carbon fiber embedded PDMS yarns. *Compos. B Eng* **2019**, *164*, 648–658. [CrossRef]
10. Toprakci, H.A.; Kalanadhabhatla, S.K.; Spontak, R.J.; Ghosh, T.K. Polymer nanocomposites containing carbon nanofibers as soft printable sensors exhibiting strain-reversible piezoresistivity. *Adv. Funct. Mater.* **2013**, *23*, 5536–5542. [CrossRef]
11. Jing, Z.; Guang-Yu, Z.; Dong-Xia, S. Review of graphene-based strain sensors. *Chin. Phys. B* **2013**, *22*, 057701.
12. Park, S.-H.; Hwang, J.; Park, G.-S.; Ha, J.-H.; Zhang, M.; Kim, D.; Yun, D.-J.; Lee, S.; Lee, S.H. Modeling the electrical resistivity of polymer composites with segregated structures. *Nat. Commun.* **2019**, *10*, 1–11. [CrossRef] [PubMed]
13. Li, J.; Ma, P.C.; Chow, W.S.; To, C.K.; Tang, B.Z.; Kim, J.K. Correlations between percolation threshold, dispersion state, and aspect ratio of carbon nanotubes. *Adv. Funct. Mater.* **2007**, *17*, 3207–3215. [CrossRef]
14. Duan, L.; Fu, S.; Deng, H.; Zhang, Q.; Wang, K.; Chen, F.; Fu, Q. The resistivity-strain behavior of conductive polymer composites: Stability and sensitivity. *J. Mater. Chem.* **2014**, *2*, 17085–17098. [CrossRef]
15. Canali, C.; Malavasi, D.; Morten, B.; Prudenziati, M.; Taroni, A. Piezoresistive effects in thick-film resistors. *J. Appl. Phys.* **1980**, *51*, 3282–3288. [CrossRef]
16. Avilés, F.; Oliva, A.I.; Ventura, G.; May-Pat, A.; Oliva-Avilés, A.I. Effect of carbon nanotube length on the piezoresistive response of poly (methyl methacrylate) nanocomposites. *Eur. Polym. J.* **2019**, *110*, 394–402. [CrossRef]

17. Avilés, F.; Oliva-Avilés, A.I.; Cen-Puc, M. Piezoresistivity, strain, and damage self-sensing of polymer composites filled with carbon nanostructures. *Adv. Eng. Mater.* **2018**, *20*, 1701159. [CrossRef]
18. Lin, L.; Liu, S.; Zhang, Q.; Li, X.; Ji, M.; Deng, H.; Fu, Q. Towards tunable sensitivity of electrical property to strain for conductive polymer composites based on thermoplastic elastomer. *ACS Appl. Mater.* **2013**, *5*, 5815–5824. [CrossRef]
19. Zhao, J.; Dai, K.; Liu, C.; Zheng, G.; Wang, B.; Liu, C.; Chen, J.; Shen, C. A comparison between strain sensing behaviors of carbon black/polypropylene and carbon nanotubes/polypropylene electrically conductive composites. *Compos. Part A Appl. Sci. Manuf.* **2013**, *48*, 129–136. [CrossRef]
20. Jin, L.; Chortos, A.; Lian, F.; Pop, E.; Linder, C.; Bao, Z.; Cai, W. Microstructural origin of resistance–strain hysteresis in carbon nanotube thin film conductors. *Proc. Natl. Acad. Sci. USA* **2018**, *115*, 1986–1991. [CrossRef]
21. Simões, S.; Viana, F.; Reis, M.A.; Vieira, M.F. Aluminum and nickel matrix composites reinforced by CNTs: Dispersion/mixture by ultrasonication. *Metals* **2017**, *7*, 279. [CrossRef]
22. Ha, J.-H.; Lee, S.-E.; Park, S.-H. Effect of Dispersion by Three-Roll Milling on Electrical Properties and Filler Length of Carbon Nanotube Composites. *Materials* **2019**, *12*, 3823. [CrossRef] [PubMed]
23. Eitan, A.; Fisher, F.; Andrews, R.; Brinson, L.C.; Schadler, L. Reinforcement mechanisms in MWCNT-filled polycarbonate. *Compos. Sci. Technol.* **2006**, *66*, 1162–1173. [CrossRef]
24. Wagner, H.D.; Lourie, O.; Feldman, Y.; Tenne, R. Stress-induced fragmentation of multiwall carbon nanotubes in a polymer matrix. *Appl. Phys. Lett.* **1998**, *72*, 188–190. [CrossRef]
25. Wan, H.; Delale, F.; Shen, L. Effect of CNT length and CNT-matrix interphase in carbon nanotube (CNT) reinforced composites. *Mech. Res. Commun.* **2005**, *32*, 481–489. [CrossRef]
26. Souri, H.; Yu, J.; Jeon, H.; Kim, J.W.; Yang, C.-M.; You, N.-H.; Yang, B.J. A theoretical study on the piezoresistive response of carbon nanotubes embedded in polymer nanocomposites in an elastic region. *Carbon* **2017**, *120*, 427–437. [CrossRef]
27. Kirkpatrick, S. Percolation and conduction. *Rev. Mod. Phys.* **1973**, *45*, 574. [CrossRef]
28. Deutscher, G. Percolation and superconductivity. In *Percolation, Localization, and Superconductivity*; Springer: Berlin/Heidelberg, Germany, 1984; pp. 95–113.
29. Bauhofer, W.; Kovacs, J.Z. A review and analysis of electrical percolation in carbon nanotube polymer composites. *Compos. Sci. Technol.* **2009**, *69*, 1486–1498. [CrossRef]
30. Dalmas, F.; Cavaillé, J.-Y.; Gauthier, C.; Chazeau, L.; Dendievel, R. Viscoelastic behavior and electrical properties of flexible nanofiber filled polymer nanocomposites. Influence of processing conditions. *Compos. Sci. Technol.* **2007**, *67*, 829–839. [CrossRef]
31. Mamunya, Y.; Boudenne, A.; Lebovka, N.; Ibos, L.; Candau, Y.; Lisunova, M. Electrical and thermophysical behaviour of PVC-MWCNT nanocomposites. *Compos. Sci. Technol.* **2008**, *68*, 1981–1988. [CrossRef]
32. Meeuw, H.; Viets, C.; Liebig, W.V.; Schulte, K.; Fiedler, B. Morphological influence of carbon nanofillers on the piezoresistive response of carbon nanoparticle/epoxy composites under mechanical load. *Eur. Polym. J.* **2016**, *85*, 198–210. [CrossRef]
33. Panozzo, F.; Zappalorto, M.; Quaresimin, M. Analytical model for the prediction of the piezoresistive behavior of CNT modified polymers. *Compos. B Eng.* **2017**, *109*, 53–63. [CrossRef]
34. Ren, X.; Chaurasia, A.K.; Oliva-Avilés, A.I.; Ku-Herrera, J.J.; Seidel, G.D.; Avilés, F. Modeling of mesoscale dispersion effect on the piezoresistivity of carbon nanotube-polymer nanocomposites via 3D computational multiscale micromechanics methods. *Smart Mater. Struct.* **2015**, *24*, 065031. [CrossRef]
35. Hu, N.; Masuda, Z.; Yan, C.; Yamamoto, G.; Fukunaga, H.; Hashida, T. The electrical properties of polymer nanocomposites with carbon nanotube fillers. *Nanotechnology* **2008**, *19*, 215701. [CrossRef]

materials

Article

Effect of Filler Alignment on Piezo-Resistive and Mechanical Properties of Carbon Nanotube Composites

Hyunwoo Kim [1,†], **Soon-Kook Hong** [2,†], **Jae-Kwan Ryu** [3] **and Sung-Hoon Park** [1,*]

1 Department of Mechanical Engineering, Soongsil University, 369 Sangdo-ro, Dongjak-Gu, Seoul 06978, Korea; hyunwoi@naver.com

2 Department of Mechanical and Naval Architectural Engineering, Naval Academy, 1 Jungwon-ro, Jinhae-gu, Changwon-si, Kyungsangnam-do 440-749, Korea; hsk753@navy.mil.kr

3 LIG Nex1, 207 Mabuk-ro, Giheung-gu, Yongin-si, Gyeonggi-do 16911, Korea; jaekwan.ryu@gmail.com

* Correspondence: leopark@ssu.ac.kr; Tel.: +82-2-828-7021

† These authors contributed equally to this work.

Received: 30 April 2020; Accepted: 4 June 2020; Published: 7 June 2020

Abstract: Highly aligned multi-walled carbon nanotube (MWCNT) polymer composites were fabricated via a roll-to-roll milling process; the alignment of the MWCNTs could be controlled by varying the speed of the rotating rolls. The effect of MWCNT alignment on the polymer matrix was morphologically observed and quantitatively characterized using polarized Raman spectroscopy. To provide a more detailed comparison, MWCNT composites with alignment in the transverse direction and random alignment were fabricated and tested. Enhanced mechanical and electrical properties were obtained for the aligned MWCNT composite, which can be attributed to the efficient electrical network and load transfer, respectively. In addition, a cyclic stretching test was conducted to evaluate the piezo-resistive characteristics of the aligned MWCNT composites. The composites with an aligned filler configuration showed an exceptionally high degree of strain sensitivity compared to the other composites.

Keywords: carbon nanotube; polymer composite; aligned MWCNT; piezo-resistive characteristics

1. Introduction

Carbon nanotubes (CNTs) have shown great potential for improving the performance of polymer matrixes when used as a filler, owing to their unique electrical [1–3], mechanical [4,5] and thermal [6–8] properties. Conductive CNT polymer composites demonstrate sensitive changes in electrical resistance under an applied external force, thus making ideal candidates for strain sensor applications [9–12]. Despite the development of various conductive composites, with carbon black and graphene as fillers [13,14], composites consisting of a one dimensional (1-D) filler with a high aspect ratio (such as CNT or silver wire) in an insulating polymer matrix have advantages regarding low electrical percolation and an efficient electrical network [15]. In previous studies, depending on the alignment direction of the CNTs, the physical properties of CNT composites, such as electrical conductivity, mechanical properties and strain-resistance sensitivity, could be changed [16–19]. Various attempts to form aligned CNTs have been successfully reported [9,10,20–24]. Kaushik et al. [10] produced an aligned CNT polycarbonate composite used as a strain sensor via a micro-injection method and observed its characteristics along the CNT-alignment direction. The direct writing technique of depositing a CNT composite on a substrate similar to a printer that extrudes ink on paper tends to align the CNTs in the extrusion direction during extrusion of the CNT composite due to the influence of shear flow [25]. Alternatively, Sohn et al. [6] fabricated aligned CNT polydimethylsiloxane (PDMS) composite films via roll-to-roll milling, discovering a phenomenon in which CNTs were partially aligned; they measured electrical conductivity along the aligned and transverse directions as a function of multi-walled CNT concentration and alignment direction.

In this study, we fabricated a highly aligned multi-walled CNT/PDMS composite film via a roll-to-roll milling process. The roll-to-roll milling method aligns the inner multi-walled carbon nanotubes (MWCNTs) by applying a circumferential shear to the uncured composite paste injected between the rolls, which rotate at different speeds; the degree of alignment can be adjusted by changing the roll speed. The resulting MWCNT composite with different degrees of alignment was compared with randomly oriented and transverse direction MWCNT composites. In order to characterize the MWCNT alignment in nanotube composites, polarized Raman spectroscopy was used for aligned MWCNTs, taking advantage of the relative intensity of the observed D and G band's sensitivity to the nanotube's orientation [5,20,21,26]. To investigate the effect of MWCNT filler alignment, mechanical tests and cyclic stretching tests were conducted for the three composite cases (aligned, random and transverse).

2. Materials and Methods

2.1. Materials

The silicone elastomer (Sylgard 184 A) and curing agent (Sylgard 184 B) were purchased from Dow Corning (Midland, MI, USA). The polydimethylsiloxane (PDMS) was formed by mixing the silicone elastomer and the curing agent at a ratio of 10:1. The MWCNT powder (JENO TUBE 6A, length = 50–150 μm, diameter = 5–7 nm) was purchased from JEIO (Incheon, Korea).

2.2. Fabrication of MWCNT/PDMS Composite

To thoroughly disperse the highly entangled MWCNTs throughout the polymer matrix, 5 wt % of MWCNT and PDMS were premixed using a paste mixer for several minutes and subsequently three-roll milled for 10 min, while decreasing the gaps between the rolls. A roll-to-roll mill (tx-2053st-sp, intec, Gyeonggi-do, Korea) was constructed to fabricate an aligned MWCNT composite film. The diameter of both rolls was 56 mm and the roll gap was adjustable in increments of 10 μm. Before the paste was injected, a film tape was attached to the roll surface so that the paste on the surface could later be easily removed. When V_1 (velocity of roll #1) was faster than V_2 (velocity of roll #2), the paste injected between the rolls was attached to the surface of roll #1, forming a film. Rolls rotating at different speeds and directions apply a shear force to the paste, aligning the MWCNTs in the circumferential direction (Figure 1). Due to the nature of the roll-to-roll method, a relatively high shear force was applied to the surface compared to the inside of the film. To minimize this, a film was fabricated by injecting a specific quantity of paste while increasing the roll interval from 40 to 80 μm. Each process lasted 5 min, allowing sufficient alignment of the MWCNTs. The irregular pattern on the film surface resulting from roll-to-roll milling was an obstacle to the accurate analysis of the MWCNT's alignment characteristics. Thus, after removing the film tape adhered to the roll surface, it was cured by applying light pressure (5 MPa) via a hot film-presser (Qmesys, Gyeonggi-do, Korea), at 150 °C for 1 h. A three-roll milled random array MWCNT/PDMS composite film and a pure PDMS film were fabricated by the same equipment under 15 MPa at 150 °C for 1 h, for use as the control group. In this study, to confirm the characteristics according to the degree of alignment of the MWCNTs, two types of composite films were fabricated with V_1:V_2 = 80:360 and 80:120 (RPM). For convenience, the ratio of V_2/V_1 is expressed as R_V (i.e., the ratio of the roll's velocity). The thicknesses of the MWCNT/PDMS composite films made of R_V = 4.5 and 1.5 were both 450 μm, and the random array MWCNT/PDMS composite film was 500 μm.

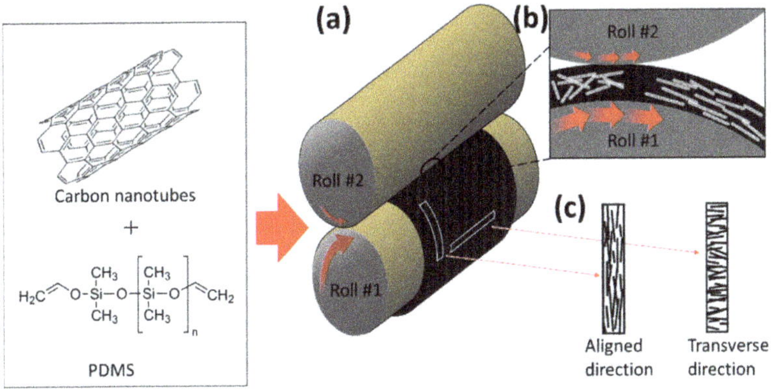

Figure 1. Fabrication scheme of the aligned multi-walled carbon nanotube (MWCNT)/PDMS composite. (**a**) Diagram of roll-to-roll milling. (**b**) The internal MWCNTs orientated in the circumferential direction, caused by shear force. (**c**) The aligned direction and transverse direction were obtained by cutting the film in the circumferential and axial directions, respectively.

2.3. Characterization

Nano/microstructure morphologies of MWCNT/PDMS composite films were characterized using SEM (GeminiSEM 300, Carl Zeiss Microscopy GmbH, Land Baden-Württemberg, Germany). The films were soaked in liquid nitrogen prior to cutting the cross-sectional areas. Polarized Raman spectroscopy, with an excitation wavelength of 514 nm, was performed to characterize the MWCNT alignment in the composite, using a Raman spectrometer (LabRam Aramis, Horiba Jobin Yvon, Kyoto, Japan). In the MWCNT-aligned complex, intensity data from the G and D peak was obtained when the MWCNT was parallel to the incident light and the scattered light was parallel (G_{II} and D_{II}) and perpendicular (G_T and D_T) to the incident light. Both results were obtained at the same location. For the random array MWCNT composite, data was collected at a fixed location in the same manner. During the tensile stretching and releasing cyclic tests, piezo-resistances were measured using a multimeter (DMM7510, Keithley, Cleveland, OH, USA) at room temperature. The tensile cyclic tests were performed at 20% of the original film length in the stretching direction, the number of cycles was 5 and the stretching and releasing speed was 20 mm/min. The length and width of the sample were 45 mm and 5 mm, respectively, and silver paste and copper tape were applied to both ends of the sample to minimize contact resistance. Mechanical properties were tested at room temperature using a universal tensile machine (DR-100, DRTECH, Seongnam, Korea) with a crosshead speed of 50 mm/min and the composite samples were prepared into dog-bone shapes.

3. Results and Discussion

After the high viscosity MWCNT/PDMS composite paste (5 wt % MWCNT) was injected between two rotating rolls, the paste mostly formed on roll #1, which had a high roll speed, as shown in Figure 1a,b. MWCNT alignment in the composite occurred in the circumferential direction due to shear forces between roll #2 and the paste interface. The shear force was not quantitatively evaluated when uniform MWCNT alignment in the composite was achieved via roll-to-roll milling. However, as shown in Figure 2, the uniformly aligned MWCNT configuration in the composite and different degrees of alignment were confirmed by the R_V values. The higher the R_V value, the higher the degree of alignment of the MWCNTs. The aligned MWCNT composite film specimens cut in the circumferential and axial direction were referred to as aligned direction and transverse direction specimens, respectively (Figure 1).

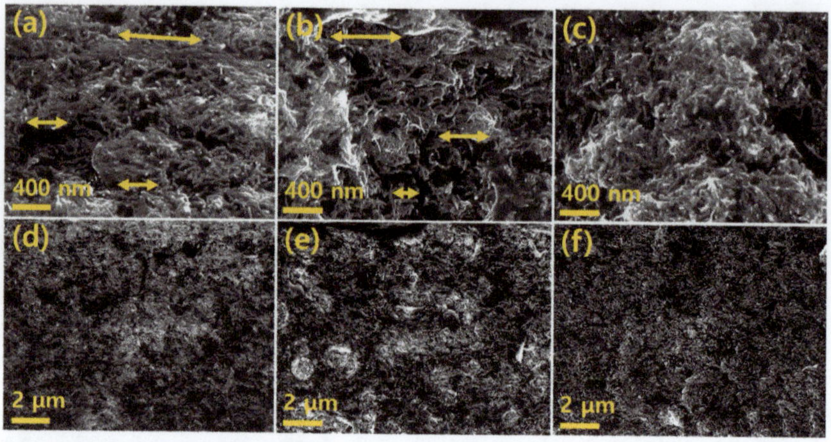

Figure 2. SEM cross-section images of: (**a**,**d**), aligned direction (R_V = 4.5). (**b**,**e**), aligned direction (R_V = 1.5). (**c**,**f**), random array MWCNT/PDMS composite film.

3.1. Morphology and Raman Spectroscopy Analysis

SEM was used to confirm the morphology of the MWCNTs and MWCNT-dispersion inside the composite through a roll-to-roll milling process. Figure 2 shows the cross-section images of a MWCNT composite film formed at: R_V = 4.5, Figure 2a,d; R_V = 1.5, Figure 2b,e and random array, Figure 2c,f. It can be seen from the image of the random array sample that the MWCNTs were uniformly dispersed throughout the composite. As shown in Figure 2a, the MWCNTs were mostly aligned in the arrow direction; there is a significant difference compared to the randomly oriented MWCNTs shown in Figure 2c. MWCNTs at R_V = 1.5 have a smaller degree of alignment compared to those at R_V = 4.5, however, the alignment degree is high in comparison with the random array. Hence, it can be seen that the degree of alignment of MWCNTs increases as the R_V value increases.

Polarized Raman spectroscopy was used to qualitatively analyze the level of alignment and the results are shown in Figure 3 and Table 1. The D band observed at 1350 cm^{-1} corresponds to disorders in the carbon atom's bonds in CNTs and the G band observed at 1590 cm^{-1} corresponds to the graphitization degree of MWCNTs [21,27]. Increments in the D_{II}/D_T and G_{II}/G_T values correspond to the enhancement of MWCNT alignment in the polymer matrix [21,26,28]. As a result, G_{II}/G_T increased from random, aligned (R_V = 1.5), and aligned (Rv = 4.5) to 1.65, 1.83 and 2.08, respectively, and D_{II}/D_T increased to 1.98, 2.10 and 2.44, respectively. As R_V was increased, the alignment of the MWCNTs increased accordingly.

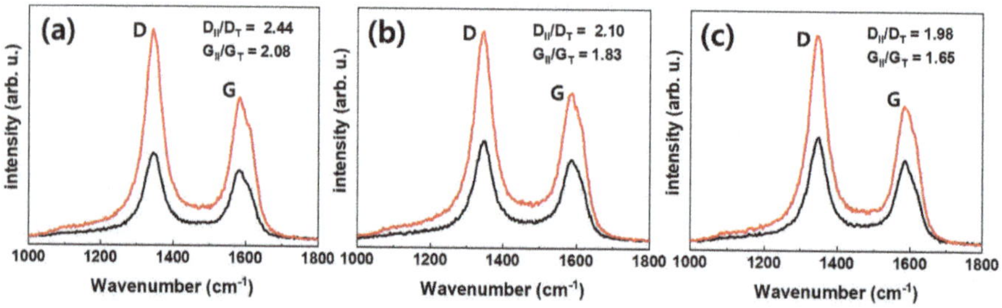

Figure 3. Raman spectra of multi-walled carbon nanotube (MWCNT)/PDMS composites; The red line measured at the incident light and the scattered light are parallel, and the black line measured at the incident light and the scattered light are perpendicular. (**a**) Aligned at R_V = 4.5, (**b**) aligned at R_V = 1.5 and (**c**) random array.

Table 1. Electrical/mechanical properties and Raman intensity ratio of the CNT/PDMS composites and pure PDMS.

Align Direction	R_V (V_1/V_2)	Electrical Conductivity [S/m]	Tensile Strength [MPa]	Young's Modulus [MPa]	Raman Intensity Ratio Parallel/Perpendicular	
					G_{II}/G_T	D_{II}/D_T
Aligned	4.5	101	9.72	7.57	2.08	2.44
	1.5	92	8.53	5.12	1.83	2.10
Transverse	4.5	57	9.16	4.00	-	-
	1.5	67	8.77	4.09	-	-
Random	-	88	7.55	4.65	1.65	1.98
Pure PDMS	-	-	6.72	1.58	-	-

3.2. Mechanical Properties

The mechanical properties depend on the load transfer capability between the MWCNTs and the matrix; an improved load transfer signifies enhanced mechanical properties. This can be achieved by aligning the CNTs [4,5,20,29]. We investigated the effects on mechanical properties of aligning MWCNTs in a polymer matrix. The analysis was conducted on the aligned MWCNT composite films, fabricated by roll-to-roll milling. The tensile stress–strain results of the MWCNT/PDMS composites along the aligned, transverse and random directions and pure PDMS are shown in Figure 4, where a load was applied until the specimen fractured. It was confirmed that the addition of 5 wt % MWCNTs to PDMS, even without alignment, resulted in an improvement to the Young's modulus and tensile strength by 194% and 12%, respectively, compared to pure PDMS. The aligned direction recorded an improvement to the Young's modulus and tensile strength by 63% and 28%, respectively, compared to the random array.

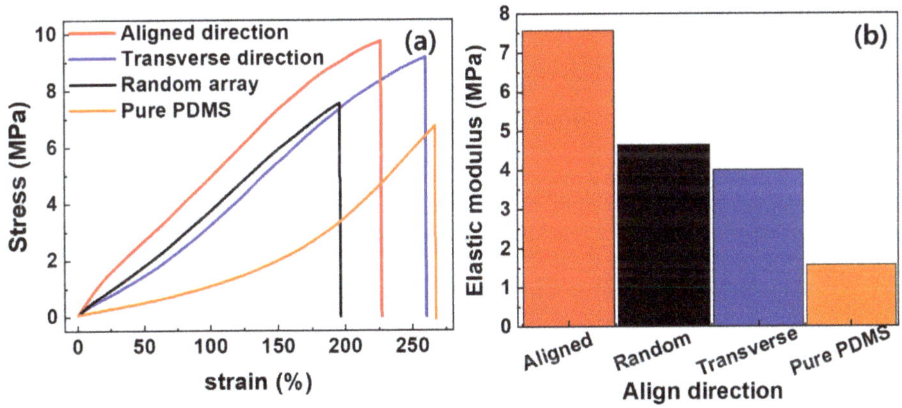

Figure 4. Mechanical properties of pure PDMS, randomly oriented, and aligned at R_V = 4.5 (aligned and transverse direction) MWCNT/PDMS composites; (**a**) stress–strain curve and (**b**) elastic modulus.

However, the transverse direction recorded a reduction in Young's modulus by 14% and an improvement to the tensile strength by 21%, compared to the random array.

The MWCNTs aligned parallel to the tensile direction showed an efficient load transfer across the MWCNT–matrix interface as the composite was stretched, with a corresponding increase in the elastic modulus. As large tensile deformation occurs, the nanotubes act as a bridge to the micro-cracks generated in the matrix, changing the crack direction and delaying fracture, thereby improving elongation [30,31]. For MWCNTs aligned perpendicular to the tensile direction, the initial tension showed poor load transfer across the MWCNT–matrix interface, however, as a larger strain was applied in the tensile direction the MWCNT also partially aligned along the tensile direction; eventually improving the extensibility [32]. Random array specimens showed higher tensile stress than transverse direction specimens, for the same strain, as more MWCNTs were parallel to the tensile direction, resulting in higher load transfer. However, the random array specimen, in which MWCNTs were entangled together, had limited alignment by tensile deformation and showed lower tensile strength than the transverse direction specimen.

3.3. Electrical Properties

Inserting a CNT filler into a polymer matrix can effectively improve conductivity [7,33]. In addition, aligned CNTs can provide higher electrical conductivity compared to randomly oriented CNTs [19]. Studies have reported that the electrical properties of CNTs improve with higher aspect ratios and content [15]. Roll-to-roll milling for the alignment of MWCNTs in a composite can be an effective method for demonstrating the degree of alignment due to shear forces. The electrical conductivity of MWCNT/PDMS composites along the aligned and transverse direction, fabricated by roll-to-roll milling, and random array MWCNT/PDMS composites are shown in Table 1. The electrical conductivity of random, aligned (R_V = 1.5) and aligned (R_V = 4.5) were 88, 92 and 101 S/m, respectively, and the random, transverse (R_V = 1.5), and transverse (R_V = 4.5) were 88, 67 and 57 S/m, respectively. In the case of the MWCNT/PDMS composite along the aligned direction, MWCNTs were mostly positioned parallel to the direction of the applied current, forming the most electrical networks, thus the electrical conductivity increased [11]. Furthermore, this tendency is clearly shown as the degree of alignment increased. Conversely, the transverse composite, in which the MWCNT was positioned perpendicular to the direction of the applied current, contained the smallest number of electrical networks; this resulted in decreased conductivity. The measured electrical conductivities are listed in Table 1.

3.4. Piezo-Resistive Properties

Piezo-resistive properties are important for obtaining information about the electrical network coupling of MWCNTs; different changes in resistance can be observed by applying an external force to the specimen. We investigated the piezo-resistance of the aligned and transverse direction composites, fabricated via a roll-to-roll milling method and random array of the MWCNT networks in a polymer matrix; this was to determine the performance as a strain sensor. Figure 5 shows the variation in resistance during five cyclic stretching tests, applying 20% tensile strain to three differently oriented specimens. After the first stretching process, $\Delta R/R_0$ was 35%, 22% and 15% in the aligned direction, transverse direction and random array, respectively. The order of these results did not match the order of electrical conductivity. To explain the resistance change during stretching, Jin et al. [34] reported that when randomly oriented CNT composite film is stretched, the CNTs reorient in the direction of stretching and slide past each other. Thus, during the tensile process it can be assumed that applying tension to the random array MWCNT/PDMS composite film partially reorients the direction of the MWCNTs in the stretching direction and creates new contact points, such that the percent of change in resistance is small. Similar to the random array sample, stretching the transverse direction composite film partially reorients the MWCNTs into the stretched direction, which induces the creation of new contacts. However, in the transverse direction composite film there are fundamentally a small number of networks and many contacts in the electrical path. Therefore, it is reasonable to assume that increasing the contact distance of MWCNTs involves a large change in resistance. Furthermore, in the process of stretching the aligned direction MWCNT composite, the specimen showed the greatest change in resistance, with the small possibility of creating new contacts as the MWCNT contact distance increased. Subsequent to the first releasing process of all specimens, $\Delta R/R_0$ decreased due to the reconnection between MWCNTs and the relocation of the electrically conductive path. It was observed that the initial values of $\Delta R/R_0$ could not be fully recovered after the first release process for each applied strain; this is due to the hysteresis effect caused by the viscoelastic matrix and permanent damage to the electrically conductive network [35]. Hysteresis confirms that the loss of the permanent network was significant in the aligned, transverse and random directions, with $\Delta R/R_0$ = 6.4%, 4.1% and 2.3%, respectively. After the first stretch-release cycle, the composite specimens demonstrated a stable sensing signal. Figure 6 shows the difference in $\Delta R/R_0$ with R_V for the specimens. For the aligned direction MWCNT/PDMS composites, the tensile-$\Delta R/R_0$ increased as R_V increased, as shown in Figure 6a. After the initial stretching cycle, the random array, R_V = 1.5 and R_V = 4.5 samples had a change in resistance of 15%, 22% and 35%, respectively. Similarly, for the transverse direction MWCNT/PDMS composites the random array, R_V = 1.5 and R_V = 4.5 samples had a change in resistance of 15%, 18% and 22%, respectively. It can be seen that the piezo-resistance increased as the degree of alignment increased. Table 2 shows the various attempts to align MWCNTs in the composite with mechanical, electrical and piezo-resistive characteristics. Results showing similar trends to our results are marked with green shaded cells. The tendency for electrical conductivity was consistent with the results we reported, while mechanical properties and resistance change during tensile test were often different. The difference could be attributed to interaction between MWCNT and the polymer matrix, the type of polymer, the content of filler and the manufacturing method.

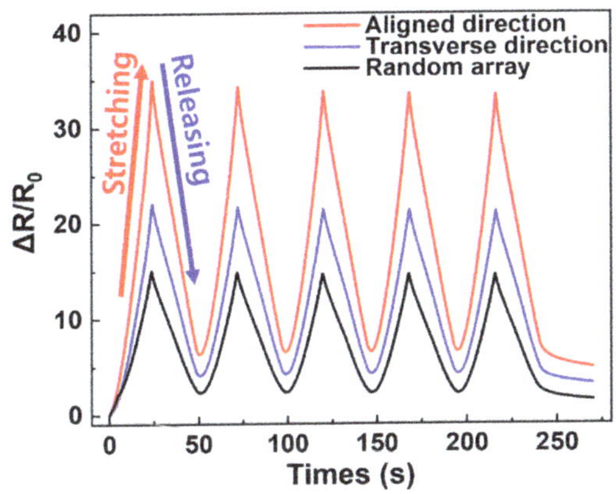

Figure 5. Piezo-resistive characteristic during stretching and releasing cyclic test from 0 to 20% strain for random, aligned at R_V = 4.5 (aligned and transverse direction) MWCNT/PDMS composites. A 40 mm strain sensor repeated at the speed of 20 mm/min.

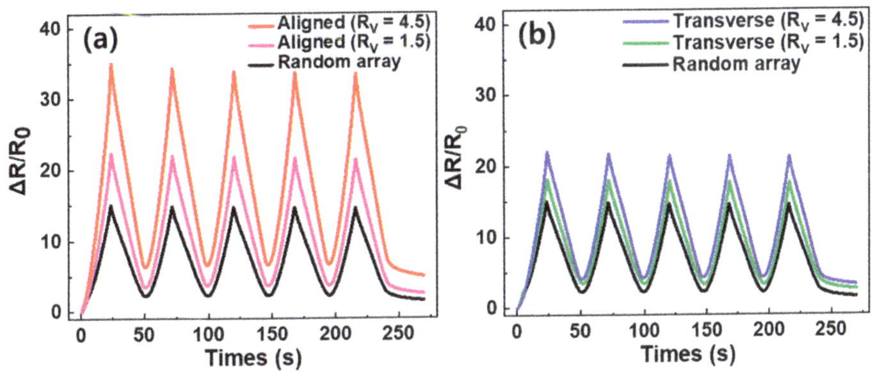

Figure 6. Comparison of CNT/PDMS composite by the degree of alignment. Cyclic test under tensile strains applied from 0 to 20% of the (**a**) aligned direction and (**b**) transverse direction.

Table 2. Electrical, mechanical properties and piezo-resistive characteristic during tensile stress of CNT/polymer based composites fabricated by various methods.

Reference and Year	Align Direction	Electrical Conductivity (S/m)	Tensile Strength (MPa)	Young's Modulus (MPa)	Resistance Change (%)	Fabrication Method	Type of Composite
Current study	Aligned direction	101 (5 wt %)	9.72	7.57	35.0 (0 to 10%)	Roll-to-roll mill, hot film-presser	5 wt % MWCNT/PDMS composite
	Transverse direction	57 (5 wt %)	9.16	4.00	21.9 (0 to 10%)		
	Random array	88 (5 wt %)	7.55	4.65	14.9 (0 to 10%)	hot film-presser	
[10], 2013	Aligned direction	R_0 = 13.9 kΩ	-	-	ΔR/ΔL = 3.65	Injection molding	5 wt % MWCNT/poly-carbonate composite
	Transverse direction	R_0 = 39.9 kΩ	-	-	ΔR/ΔL = 6.5		
[21], 2018	Aligned direction	3×10^{-4}	-	5500	-	DC electrical fields applied	0.5 wt % MWCNT/epoxy resin composite
	Transverse direction	9×10^{-5}	-	-	-		
	Random array	6×10^{-5}	-	3900	-		
[11], 2016	Aligned direction	8×10^{-5}	34	1200	-	DC electrical fields applied	MWCNT/poly vinylidene fluoride composite
	Transverse direction	1.5×10^{-5}	39	1210	-		
	Random array	1.8×10^{-5}	26	950	-		
[4], 2015	Aligned direction	5.25×10^{-3}	72	1680	-	DC electrical fields applied	0.5 wt % MWCNT/chito-san composite
	Transverse direction	4.57×10^{-8}	50	1620	-		
	Random array	4.77×10^{-3}	68	1640	-		
[12], 2018	Aligned direction	R_0 = 4.96 kΩ	9.5	4.7	9	CNT film on the PDMS	
	Transverse direction	R_0 = 69.6 kΩ	7.5	4.4	1		

4. Conclusions

We confirmed that the degree and direction of alignment MWCNTs in the polymer composite affects the composite's mechanical, electrical and electro-mechanical properties. It has been demonstrated that roll-to-roll milling can significantly influence the alignment of MWCNTs in a polymer matrix. SEM images, D_{II}/D_T and G_{II}/G_T from polarized Raman spectroscopy showed that as R_V increased, the alignment of the MWCNTs increased, which was attributed to the applied shear forces. The electrical conductivity, Young's modulus, tensile strength and strain sensitivity were highest for the aligned direction MWCNT/PDMS composite due to the effective load transfer and conductive network of MWCNTs. On the contrary, the electrical conductivity and Young's modulus were lowest for the transverse direction MWCNT/PDMS composite.

Author Contributions: Conceptualization, S.-H.P.; methodology, H.K. and S.-K.H.; formal analysis, S.-H.P.; resources, H.K. and J.-K.R.; writing—original draft preparation, H.K.; writing—review and editing, S.-K.H.; supervision, S.-H.P.; project administration, S.-K.H.; funding acquisition, S.-H.P. All authors have read and agreed to the published version of the manuscript.

Funding: This work was supported by the National Research Foundation of Korea (NRF) grant funded by the Korea government (MSIT No. 2020R1A2C1013489). This research is part of the result carried out by support of Lig Nex 1.

Conflicts of Interest: The authors declare no conflict of interest.

References

1. Chu, K.; Kim, D.; Sohn, Y.; Lee, S.; Moon, C.; Park, S. Electrical and Thermal Properties of Carbon-Nanotube Composite for Flexible Electric Heating-Unit Applications. *IEEE Electron. Device Lett.* **2013**, *34*, 668–670. [CrossRef]
2. Chu, K.; Park, S.-H. Electrical heating behavior of flexible carbon nanotube composites with different aspect ratios. *J. Ind. Eng. Chem.* **2016**, *35*, 195–198. [CrossRef]
3. Chu, K.; Yun, D.-J.; Kim, D.; Park, H.; Park, S.-H. Study of electric heating effects on carbon nanotube polymer composites. *Org. Electron.* **2014**, *15*, 2734–2741. [CrossRef]
4. Gupta, P.; Sharan, S.; Roy, P.; Lahiri, D. Aligned carbon nanotube reinforced polymeric scaffolds with electrical cues for neural tissue regeneration. *Carbon* **2015**, *95*, 715–724. [CrossRef]
5. Mikhalchan, A.; Gspann, T.; Windle, A. Aligned carbon nanotube–epoxy composites: The effect of nanotube organization on strength, stiffness, and toughness. *J. Mater. Sci.* **2016**, *51*, 10005–10025. [CrossRef]
6. Sohn, Y.; Kim, D.E.; Park, S.T.; Lee, S.-E. Seamless Tube-Type Heater with Uniform Thickness and Temperature Distribution Based on Carbon Nanotubes Aligned by Circumferential Shearing. *Materials* **2019**, *12*, 3283. [CrossRef]
7. Bauhofer, W.; Kovacs, J.Z. A review and analysis of electrical percolation in carbon nanotube polymer composites. *Compos. Sci. Technol.* **2009**, *69*, 1486–1498. [CrossRef]
8. Ha, J.H.; Chu, K.; Park, S.H. Electrical Properties of the Carbon-Nanotube Composites Film Under Extreme Temperature Condition. *J. Nanosci. Nanotechnol.* **2019**, *19*, 1682–1685. [CrossRef]
9. Li, A.; Bogdanovich, A.E.; Bradford, P.D. Aligned carbon nanotube sheet piezoresistive strain sensors. *Smart Mater. Struct.* **2015**, *24*, 095004. [CrossRef]
10. Parmar, K.; Mahmoodi, M.; Park, C.; Park, S.S. Effect of CNT alignment on the strain sensing capability of carbon nanotube composites. *Smart Mater. Struct.* **2013**, *22*, 075006. [CrossRef]
11. Gupta, P.; Rajput, M.; Singla, N.; Kumar, V.; Lahiri, D. Electric field and current assisted alignment of CNT inside polymer matrix and its effects on electrical and mechanical properties. *Polymer* **2016**, *89*, 119–127. [CrossRef]
12. Ma, L.; Yang, W.; Wang, Y.; Chen, H.; Xing, Y.; Wang, J. Multi-dimensional strain sensor based on carbon nanotube film with aligned conductive networks. *Compos. Sci. Technol.* **2018**, *165*, 190–197. [CrossRef]
13. Shintake, J.; Piskarev, E.; Jeong, S.H.; Floreano, D. Ultrastretchable Strain Sensors Using Carbon Black-Filled Elastomer Composites and Comparison of Capacitive Versus Resistive Sensors. *Adv. Mater. Technol.* **2018**, *3*, 1700284. [CrossRef]
14. Wang, Y.; Wang, L.; Yang, T.; Li, X.; Zang, X.; Zhu, M.; Wang, K.; Wu, D.; Zhu, H. Wearable and Highly Sensitive Graphene Strain Sensors for Human Motion Monitoring. *Adv. Funct. Mater.* **2014**, *24*, 4666–4670. [CrossRef]
15. Ha, J.-H.; Lee, S.-E.; Park, S.-H. Effect of Dispersion by Three-Roll Milling on Electrical Properties and Filler Length of Carbon Nanotube Composites. *Materials* **2019**, *12*, 3823. [CrossRef] [PubMed]
16. Junyong, L.; Miao, L.; Bermak, A.; Yi-Kuen, L. Study of piezoresistance effect of carbon nanotube-PDMS composite materials for nanosensors. In Proceedings of the 2007 7th IEEE Conference on Nanotechnology (IEEE NANO), Hong Kong, China, 2–5 August 2007; pp. 1240–1243.
17. Liu, S.; Tian, M.; Zhang, L.; Lu, Y.; Chan, T.W.; Ning, N. Tailoring dielectric properties of polymer composites by controlling alignment of carbon nanotubes. *J. Mater. Sci.* **2016**, *51*, 2616–2626. [CrossRef]
18. Larijani, M.M.; Khamse, E.J.; Asadollahi, Z.; Asadi, M. Effect of aligned carbon nanotubes on electrical conductivity behaviour in polycarbonate matrix. *Bull. Mater. Sci.* **2012**, *35*, 305–311. [CrossRef]
19. Lanticse, L.J.; Tanabe, Y.; Matsui, K.; Kaburagi, Y.; Suda, K.; Hoteida, M.; Endo, M.; Yasuda, E. Shear-induced preferential alignment of carbon nanotubes resulted in anisotropic electrical conductivity of polymer composites. *Carbon* **2006**, *44*, 3078–3086. [CrossRef]
20. Khan, S.U.; Pothnis, J.R.; Kim, J.-K. Effects of carbon nanotube alignment on electrical and mechanical properties of epoxy nanocomposites. *Compos. Part A Appl. Sci. Manuf.* **2013**, *49*, 26–34. [CrossRef]
21. Amani, A.; Hashemi, S.A.; Mousavi, S.M.; Pouya, H.; Vojood, A. Electric Field Induced Alignment of Carbon Nanotubes: Methodology and Outcomes. In *Carbon Nanotubes-Recent Progress*; IntechOpen: London, UK, 2018; pp. 71–88.

22. Goh, G.L.; Agarwala, S.; Yeong, W.Y. Aerosol-Jet-Printed Preferentially Aligned Carbon Nanotube Twin-Lines for Printed Electronics. *ACS Appl. Mater. Interfaces* **2019**, *11*, 43719–43730. [CrossRef]
23. Goh, G.L.; Saengchairat, N.; Agarwala, S.; Yeong, W.Y.; Tran, T. Sessile droplets containing carbon nanotubes: A study of evaporation dynamics and CNT alignment for printed electronics. *Nanoscale* **2019**, *11*, 10603–10614. [CrossRef] [PubMed]
24. Herren, B.; Gu, T.; Tang, Q.; Saha, M.; Liu, Y. 3D Printing and Stretching Effects on Alignment Microstructure in PDMS/CNT Nanocomposites. In Proceedings of the ASME 2019 International Mechanical Engineering Congress and Exposition, Salt Lake City, UT, USA, 11–14 November 2019.
25. Agarwala, S.; Goh, G.L.; Goh, G.D.; Dikshit, V.; Yeong, W.Y. Chapter 10—3D and 4D printing of polymer/CNTs-based conductive composites. In *3D and 4D Printing of Polymer Nanocomposite Materials*; Sadasivuni, K.K., Deshmukh, K., Almaadeed, M.A., Eds.; Elsevier: Amsterdam, The Netherlands, 2020; pp. 297–324.
26. Abbasi, S.; Carreau, P.J.; Derdouri, A. Flow induced orientation of multiwalled carbon nanotubes in polycarbonate nanocomposites: Rheology, conductivity and mechanical properties. *Polymer* **2010**, *51*, 922–935. [CrossRef]
27. Abu Bakar, S.; Park, J. Alignment of multi-walled carbon nanotubes in a polyethylene matrix by extrusion shear flow: Mechanical properties enhancement. *J. Compos. Mater.* **2010**, *45*, 931–941. [CrossRef]
28. Han, B.; Xue, X.; Xu, Y.; Zhao, Z.; Guo, E.; Liu, C.; Luo, L.; Hou, H. Preparation of carbon nanotube film with high alignment and elevated density. *Carbon* **2017**, *122*, 496–503. [CrossRef]
29. Zhou, Z.; Wang, X.; Faraji, S.; Bradford, P.D.; Li, Q.; Zhu, Y. Mechanical and electrical properties of aligned carbon nanotube/carbon matrix composites. *Carbon* **2014**, *75*, 307–313. [CrossRef]
30. Li, H.; Dai, X.; Zhao, L.; Li, B.; Wang, H.; Liang, C.; Fan, J. Microstructure and properties of carbon nanotubes-reinforced magnesium matrix composites fabricated via novel in situ synthesis process. *J. Alloys Compd.* **2019**, *785*, 146–155. [CrossRef]
31. Paramsothy, M.; Chan, J.; Kwok, R.; Gupta, M. Addition of CNTs to enhance tensile/compressive response of magnesium alloy ZK60A. *Compos. Part A Appl. Sci. Manuf.* **2011**, *42*, 180–188. [CrossRef]
32. Yu, Y.; Luo, Y.; Guo, A.; Yan, L.; Chun, O.; Jiang, K.; Li, Q.; Fan, S.; Wang, J. Flexible and Transparent Strain Sensors Based on Super-aligned Carbon Nanotube Films. *Nanoscale* **2017**, *9*, 6716–6723. [CrossRef]
33. Gommans, H.H.; Alldredge, J.W.; Tashiro, H.; Park, J.; Magnuson, J.; Rinzler, A.G. Fibers of aligned single-walled carbon nanotubes: Polarized Raman spectroscopy. *J. Appl. Phys.* **2000**, *88*, 2509–2514. [CrossRef]
34. Jin, L.; Chortos, A.; Lian, F.; Pop, E.; Linder, C.; Bao, Z.; Cai, W. Microstructural origin of resistance-strain hysteresis in carbon nanotube thin film conductors. *Proc. Natl. Acad. Sci. USA* **2018**, *115*, 1986–1991. [CrossRef]
35. Duan, L.; Fu, S.; Deng, H.; Zhang, Q.; Wang, K.; Chen, F. The resistivity-strain behavior of conductive polymer composites: Stability and sensitivity. *J. Mater. Chem. A* **2014**, *2*, 17085–17098. [CrossRef]

materials

MDPI

Article

Seamless Tube-Type Heater with Uniform Thickness and Temperature Distribution Based on Carbon Nanotubes Aligned by Circumferential Shearing

Yoonchul Sohn [1], Dongearn Kim [2,†], Sung-Hoon Park [3,*] and Sang-Eui Lee [4,*,†]

1. Department of Welding & Joining Science Engineering, Chosun University, 309 Pilmun-daero, Dong-gu, Gwangju 61452, Korea; yoonchul.son@chosun.ac.kr
2. Molds & Dies Technology Group, Korea Institute of Industrial Technology, 7-47 Songdo-dong, Yeonsoo-gu, Incheon 21999, Korea; kdu0517@kitech.re.kr
3. Department of Mechanical Engineering, Soongsil University, 369 Sangdo-ro, Donjak-gu, Seoul 06978, Korea
4. Department of Mechanical Engineering, Inha University, Inha-ro 100, Michuhol-gu, Incheon 22212, Korea

* Correspondence: selee@inha.ac.kr (S.-E.L.); leopark@ssu.ac.kr (S.-H.P.); Tel.: +82-32-860-7377 (S.-E.L.); +82-2-828-7021 (S.-H.P.)

† Materials Research Center, Samsung Advanced Institute of Technology, Samsung-ro 130, Suwon 16678, Korea (previous workplace).

Received: 12 September 2019; Accepted: 4 October 2019; Published: 9 October 2019

Abstract: The uniform temperature distribution, one of the requirements for long-term durability, is essential for composite heaters. An analytical model for temperature distribution of a tube-type heater was derived, and it revealed that thickness uniformity is one order more important than intrinsic material properties such as density, heat capacity, and electrical conductivity of the heating tube. We introduced a circumferential shearing process to fabricate a flexible, seamless tube-type heating layer of carbon nanotube/silicone rubber composite with outstanding uniform distribution of thickness and temperature, which may be attributed to a shorter characteristic dimension in the circumferential direction than in the axial direction. The temperature uniformity was experimentally verified at various temperatures under heating. The difference in measured thickness and temperature in circumferential direction was within ±1.3~3.0% (for t_{avg} = 352.7 μm) and ±1.1% (for T_{avg} = 138.8 °C), respectively, all over the heating tube. Therefore, the circumferential shearing process can be effective for cylindrical heaters, like a heating layer of a laser printer, which fuse toners onto papers during printing.

Keywords: carbon nanotubes; circumferential shearing; alignment; electrical conductivity

1. Introduction

Carbon nanotube (CNT) composites have been extensively investigated to make use of unique characteristics of CNT such as high electrical and thermal conductivity, robust mechanical property, and fast heating performance. Electrically conductive composites were first demonstrated, taking advantages of the high aspect ratio of the CNTs to form a percolation network [1]. For load-bearing applications, CNT composites with polymers or precursor resins can increase mechanical properties such as stiffness, strength, and toughness [2]. CNT composites are also used to enhance physical properties of carbon fiber composites, which can be primary materials for strong, lightweight wind turbine blades and hulls for maritime security boats [2,3]. Another application is a multifunctional coating material such as multiwalled nanotube (MWNT)-containing paints, reducing biofouling of ship hulls by preventing attachment of algae and barnacles [4]. In lithium ion batteries, small amounts of MWNT powder are blended with active materials and a polymer binder in $LiCoO_2$ cathodes and graphite anodes, where CNTs provide both enhanced electrical connectivity and mechanical integrity, leading to an increase in rate capability and life cycles [4,5].

Among the multifunctional applications, CNT composites have also been widely studied in heat-related applications such as patternable micro heaters, temperature sensors, heating glasses for vehicles, thermoelectric devices, water heaters, and flexible de-icing units [6–9]. By the electric Joule heating (resistive heating) of conducting composites, electrical energy can be converted into heat energy quantitatively. In general, composites that exhibit high conductivity and low heat capacity can be ideal for rapid heating applications. Therefore, CNTs can be attractive filler materials for resistive heating with their high aspect ratios coupled with high electrical and thermal conductivities, as high as 106 S/m and 6600 W/m·K, respectively [10–12]. One of the practical applications for CNT resistive heaters is a laser printer fuser [13]. The printer fuser heats up and melts toners electrostatically-attached on a paper so that the toners can be fused on the surface of the paper, while the paper passes between a heating roller and a pressing roller. In the heating roller, a conductive composite as heating element is placed on a polymer support, e.g., polyimide (PI) tube, and a protective and toner-releasing layer such as perfluoroalkoxy alkanes (PFA), as shown in Figure 1a. Coating conductive composites on the PI tube could be formed by extrusion process [13]. However, irregular thickness distribution of the conducting composites through circumferential and axial directions of the roller is often observed, as shown in Figure 1b, which may be attributed to two main causes; high viscosity (or stiffness) of CNT composites, and misalignment between the PI tube axis (or extruded core axis) and the extrusion die axis.

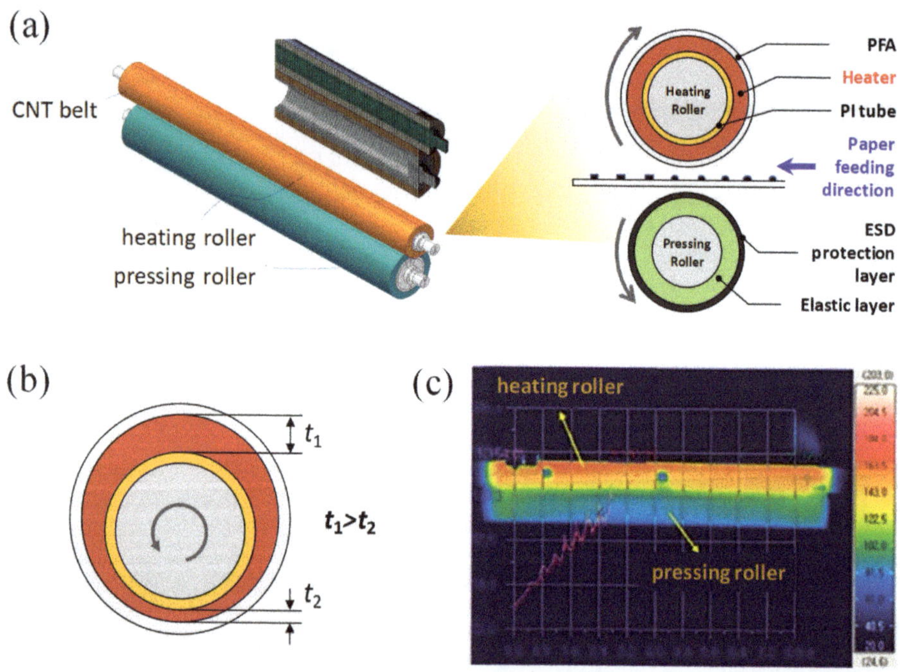

Figure 1. (**a**) Schematic of laser printer fusing element, (**b**) thickness discrepancy of multiwalled nanotube (MWNT)/rubber heater layer in circumferential direction caused by misalignment of the roller axis to extrusion axis, and (**c**) extruded heating roller showing temperature irregularity in the circumferential direction.

To obtain uniform thickness in axial and circumferential directions, the alignment between the two axes should be guaranteed. Otherwise, the uneven thickness of the heating element could result in fatal malfunctions by irregular temperature distribution during heating process, as shown in Figure 1c.

The problem can be circumvented by using a roll coating process, ensuring uniform thickness of composite materials, leading to enhancement in process repeatability, because the characteristic length of the roll coating process, the diameter of the rollers, is lower than that of the extrusion process, the length of the roller.

In this study, uniform temperature distribution along the circumferential direction, as well as in the axial direction, was achieved by applying circumferential shearing with a two roller system. Viscous pastes were able to be coated on one of the two rollers with a higher rotational speed. The MWNTs were observed to be aligned in the circumferential direction in the circumferential shearing, while the nanofillers are aligned in axial direction in the extrusion process. Fast heating performance was also demonstrated with uniform temperature distribution along both circumferential and axial directions.

2. Materials and Methods

2.1. Materials

MWNTs, purchased from Hanhwa Chemicals (Ulsan, South Korea), were 100–200 μm and 10–20 nm in length and diameter. Silicone rubber used as polymer matrix was Sylgard 184, a type of polydimethylsiloxane (PDMS), purchased from Dow Corning (Midland, MI, USA). The combination of the two materials was chosen for flexibility and elastomeric response for efficient toner fusing during printing [13], and also for material stability at least up to 230~250 °C [14–16].

2.2. Fabrication of MWNT/Silicone Composite and Printer Fuser

For the dispersion of MWNTs into PDMS, a two-step process was applied; a planetary milling as preliminary first dispersion process, and then a three-roll milling (TRM) as primary dispersion process. At first, MWNTs and PDMS were weighed and then manually mixed. The mixing ratio of the elastomer base and the curing agent of PDMS was 10:1 by weight. The mixture was planetary-milled and then three-roll milled.

Effects of the concentration of MWNTs on electrical conductivities of the MWNT/PDMS composites were evaluated in order to satisfy required axial electrical conductivity of the seamless composite heating tube for a given thickness. In the evaluation, the volume fraction of MWNT and PDMS were calculated by assuming their densities to be 1.85 g/cm^3 and 1.03 g/cm^3, respectively.

A circumferential shearing machine was designed to coat the MWNT/PDMS pastes onto a hollow cylindrical PI tube, as shown in Figure 2 and Figure S1. The diameter and the length of the rolls were 24 mm and 300 mm, respectively. The 50 μm thick polyimide tube was placed onto the 1st roller. The 1st and the 2nd roller angular speeds, v_1 and v_2, were set up to 300 and 100 rpm, respectively, based on the evaluation of film thickness uniformity, which is shown in Figure S1. Due to the speed difference of the two rollers, the MWNT pastes fed between the two rollers (Figure 2a) were rolled up onto the 1st roller at the higher rotational speed (Figure 2b), and then a uniform composite film on the 1st roller was obtained by separating the 2nd roller from the 1st one (Figure 2c).

In order to build up a printing fuser, metallic electrodes were deposited on the ends of the heating element to apply voltage, and then the MWNT/PDMS heating tube was covered together with a toner release layer of perfluoroalkoxy (PFA) tube, which was 50 μm in thickness, as shown in Figure 2d. The assembled printing fuser was cured at 100 °C at 30 min. and then 180 °C at 2 h. The exposed region of the electrode had a good contact with terminals for power supply, even during roller rotation. A pressing roller was assembled with the heating roller to supply sufficient both heat and pressure to toners on papers, as shown in Figure 1a.

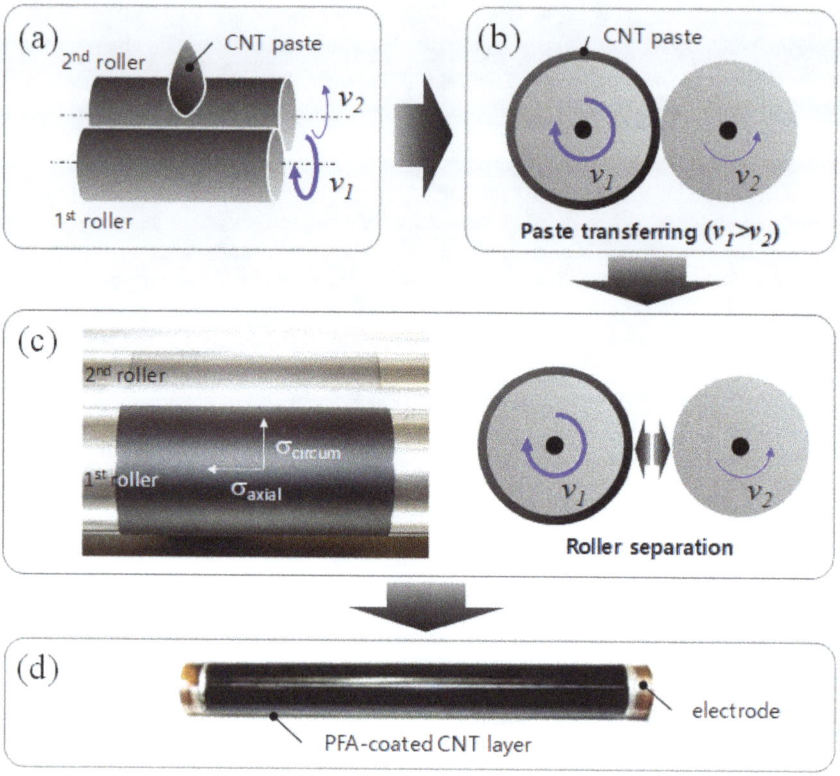

Figure 2. Circumferential shearing process for uniform coating on a cylindrical polyimide (PI) tube. (**a**) MWNT paste feeding, (**b**) paste transferring on a roller with a higher speed, (**c**) roller separation, and (**d**) MWNT heating element with a perfluoroalkoxy alkanes (PFA) layer and electrodes.

2.3. Characterization

Microstructures of MWNT/PDMS composites on the PI tube were observed by scanning electron microscope (SEM) (model: UHR-SEM S-5500, Hitachi, Tokyo, Japan). Electrical conductivity of MWNT/silicone composites was measured according to IEC Standard 93 method (VDE 0303, Part 30) using two-point source meter. Polarized Raman analysis with an excitation wavelength of 532 nm was made to evaluate degree of alignment the MWNTs in the axial and circumferential direction of heating tubes. For the observation of electric heating characteristics of the seamless heating tube, direct current (DC) was applied to the printer fuser with a power supply (NI PXI-1033, National Instruments, Austin, TX, USA). Temperature distribution over the fuser was recorded and imaged by an infrared (IR) camera (Advanced Thermo, TVS-500, Tokyo, Japan).

3. Results and Discussion

3.1. MWNT/PDMS Pastes

Carbon nanotubes used in this study are shown in Figure 3a. As-received MWNTs consisted of MWNT bundles. Diameters of the bundles were in scales of several micrometers, while some of them were over 10 μm. In a close observation of MWNT bundles, the MWNTs are entangled with each other from van der Waals interaction. The conductive fillers, MWNTs were dispersed in the elastomer, PDMS, with the two-step dispersing process using a planetary mixer and three roll mill (TRM).

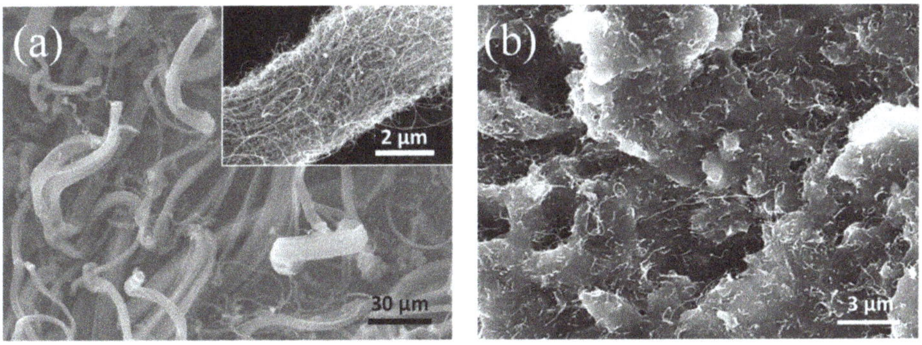

Figure 3. SEM micrographs of (**a**) as-received MWNTs, and (**b**) MWNT/PDMS composite.

For the TRM process, the repetition number of the TRM process and the pressure between the rolls were varied to obtain maximum conductivity of the composites by optimizing dispersion of the MWNTs in the silicone rubber [16]. Electrical conductivity of MWNT/silicone composites increased as the number of TRM increased and reached the maximum value after six times of repetition in our previous study [16]. Therefore, the composites which had the same TRM processing times were used for the fabrication of MWNT/PDMS paste. The MWNTs are observed to be in excellent degree of dispersion in the polymer matrix, as presented in Figure 3b.

3.2. Analytic Modelling for Temperature Uniformity of Tube-Type Heater

As shown in Figure 1a, the heating roller consists of a conductive heating element between the PI tube and PFA protective layer. To estimate the effects of thickness and electrical conductivity of the heating element on temperature uniformity under an applied voltage, an analytic model was derived for the heating tube, as presented in Figure 4. Axial and circumferential directions of the tube are denoted as z and θ, respectively. Thickness and outer radius of the heating element are denoted with t and R_o, respectively.

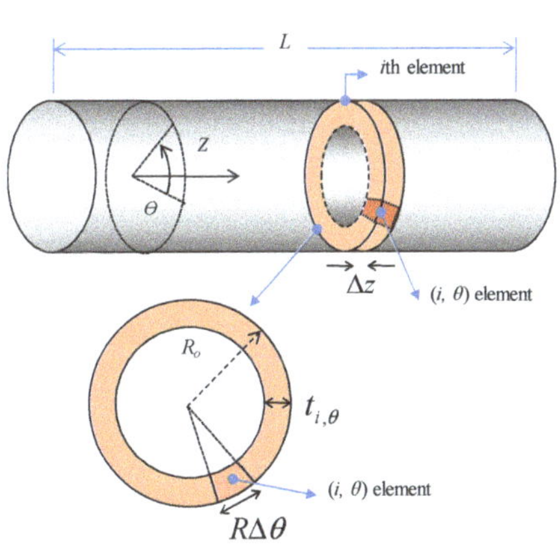

Figure 4. A schematic of tube-type heater for analytical model of temperature distribution.

There are assumptions for this model:

1. Heat loss and heat transfer on tube surface are assumed to be disregarded.
2. MWNTs are evenly dispersed, and the degree of MWNT alignment is uniform in the composite tube, which means that electrical conductivity of the (i,θ) element, $\sigma_{i,\theta\text{-}A} = \sigma_A$, and $\sigma_{i,\theta\text{-}C} = \sigma_C$, where A and C stand for axial and circumferential, respectively.

3. Since the electrodes are located at both ends of the heating tube, the total electrical current (I) flows through the axial direction. Based on that, the current (I_θ) of an axial element for the tube length (L) and a circumferential length ($R_o \cdot \Delta\theta$) are assumed to be the same, that is, $I_\theta = \Sigma_i (I_{i,\theta}) = I/(\Delta\theta/2\pi)$, where $I_{i,\theta}$ is the current flowing in the (i,θ) element in the axial direction. This assumption is reasonable in case the average values of thickness and electrical conductivity of the axial elements are approximately same.
4. The i-th section (circumferential element) is electrically connected in series, and the electrical current running through the i-th section (I_i) is constant, that is, $I_i = \Sigma_\theta (I_{i,\theta}) = I$.
5. (i,θ) elements with a cross sectional area of $t_{i,\theta} \times R_o\Delta\theta$ are electrically connected in parallel, and each element experiences the same voltage drop along Δz, the length of the ith section, that is, $V_{i,\theta} = V_i$, where V_i is the voltage drop in the i-th section.

Electrical resistance of the (i,θ) element is expressed in Equation (1).

$$r_{i,\theta} = \frac{\Delta z}{R_o \Delta\theta \cdot t_{i,\theta}} \cdot \frac{1}{\sigma_{i,\theta}} \tag{1}$$

From energy conservation, thermal energy of the heating tube is the same as electrical energy applied. $\sigma_{i,\theta}$ is electrical conductivity of (i,θ) element.

$$m_i c_i \Delta T_{i,\theta} = \frac{(V_{i,\theta})^2}{r_{i,\theta}} t \tag{2}$$

Equation (2) can be rewritten, together with Equation (1).

$$\left(\rho_{i,\theta} \cdot t_{i,\theta} \cdot R_o \Delta\theta \cdot \Delta z\right) c_{i,\theta} \Delta T_{i,\theta} = \frac{(I_{i,\theta} \cdot r_{i,\theta})^2}{r_{i,\theta}} t \tag{3}$$

where m, c, and ρ represent mass, heat capacity, and density of the heating tube, and t is the time during which the electric power is supplied. Based on the 2nd assumption, $\rho_{i,\theta} = \rho$ and $c_{i,\theta} = c$, respectively. The heating rate or thermal response, $\Delta T_{i,\theta}/t$, can be expressed in Equation (4).

$$\frac{\Delta T_{i,\theta}}{t} = \frac{I_{i,\theta}{}^2}{\rho \cdot c \cdot (R_0 \Delta\theta)^2} \cdot \frac{\sigma_{i,\theta}}{(t_{i,\theta})^2} \tag{4}$$

Therefore, the temperature at a specific point (i,θ) of the heating tube depends not only on intrinsic material properties of MWNT/PDMS composite such as electrical conductivity, density, and heat capacity, but also on extrinsic parameters such as thickness and diameter of the heating tube. The temperature is linearly dependent on the former, while it is dependent on the square of the latter geometric variables, revealing that the latter is more dominant for temperature uniformity and heating rate. A process conducted to circumferential direction is expected to have an advantage on easily controlling the geometrical parameters compared to those done to axial direction like extrusion process, because a characteristic dimension can be short in the circumferentially-controlled process, that is, circumferential length, $2\pi R_o$, is shorter than axial tube length, L. Therefore, the circumferential shear rolling process was adopted in this study.

In addition, in terms of material properties for the heating element, low heat capacity and high electrical conductivity are beneficial for fast heating. Therefore, carbon nanotubes can be the best candidate with such low specific heat of 0.75 J/g·K for MWNT [17] and high conductivity of 10^4~10^5 S/m [18].

3.3. *Printer Fuser Fabricated by Circumferential Shearing Process*

A PI tube with copper electrodes at the tube edges was inserted on a faster roller before the paste was fed. In this study, the rolling speed ratio of the 1st and 2nd rollers varied from 1.1 to 4.5, and the optimum ratio, 3.0, was set up, on the basis of film thickness uniformity (see Supplementary Materials). When the ratio was too low and high, irregular circumferential patterns with thickness differences on composite surfaces were formed, while there was a uniform surface in the optimized condition, as shown in the supporting information. After the coating process of MWNT/PDMS composite was completed, a PFA protective film was prepared on the heating tube. The final structure of the printer fuser consists of PI/heating layer/PFA with the copper electrodes on both ends of the tube, as shown in Figure 2d. For the secure contact of the electrodes and the MWNT/SR heating layer, a silver paste was applied at the edge of the composite.

The MWNTs are aligned in the circumferential direction during the shearing process, because the MWNT paste was sheared to the circumferential flow direction. On the other hand, the nanotubes are preferentially aligned in the axial direction in the extrusion process for the same reason. The filler alignment during the coating process is of great importance, since it strongly affects the electrical conductivity of the composite between electrodes, determining the total resistance of composite heaters. Equation (4) reveals that the resistance of the heating tube depends more on thickness difference than on the electrical conductivity, which is a main reason why the circumferential shearing process was introduced in this study.

Figure 5 shows the thickness variation of the tube-type heating tube in both the axial and circumferential directions, along which 11 and 6 samples were prepared, respectively, as shown in Figure 5a.

For a given axial position, the maximum thickness differences ($|t_{max} - t_{min}|_{max}$) are within ±1.3~3.0% for the average temperature in circumferential direction. For the overall average thickness of 357.9 μm (t_{avg_all}) over the whole area of the heating tube, all the measured thicknesses lay within 339~374 μm (±5% of the average), as shown in Figure 5b,c. In the axial direction of the tube, the thicknesses around the center of the tube were higher than those near the edges, as shown in Figure 5b. This may be attributed to the feeding position of the MWNT paste, as it was fed at the center of the PI tube.

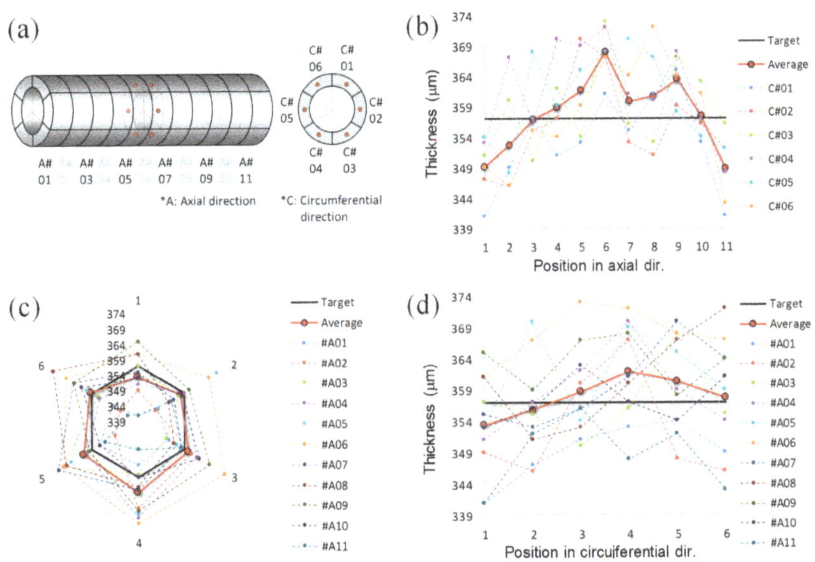

Figure 5. Analysis of thickness distribution. (**a**) denotation of subsection of composite tube, and measured thicknesses in the axial direction (**b**), and in the circumferential direction (**c**,**d**).

In the circumferential direction, the average thicknesses in the axial direction for a given circumferential position range between 353.4 μm (C#01) to 361.9 μm (C#04), close to t_{avg_all}, as shown in Figure 5c,d. Based on the analytical model for temperature uniformity in Equation (3), the outstanding uniform thickness in the axial and circumferential directions can guarantee temperature uniformity on the seamless heating tube in both directions, together with the uniform dispersion of MWNTs and the even shear rate along axial direction.

In addition, the thickness discrepancy between the center and the edges of the heating tube can be expected to be further minimized by adjusting the number of feeding positions and the amount of the MWNT paste.

3.4. Anisotropic Characteristics of Tube-Type Heating Element

Electrical conductivity of the two-roller processed MWNT/PDMS printer fuser was measured as a function of MWNT concentration and alignment direction, as presented in Figure 6a. During the two-roller coating process, MWNTs are aligned in the circumferential direction of the heating tube. Therefore, in all the specimens with varying MWNT contents, electrical conductivities in the circumferential direction are always higher than that in the axial direction.

Ratios of electrical conductivities in the circumferential to axial directions are presented in Figure 6b. With a very small amount of MWNTs, 0.1 wt%, the anisotropy of electrical conductivity was as high as ~57. It may be attributed to the fact that the MWNT content was so low that they could not be connect with each other along the axial direction, while connected along preferential alignment (circumferential) direction [18]. With more than 1 wt% of MWNTs, the anisotropy varied between 2.7 and 1.9. The ratio gradually decreased with increasing MWNT contents, as shown in Figure 6b.

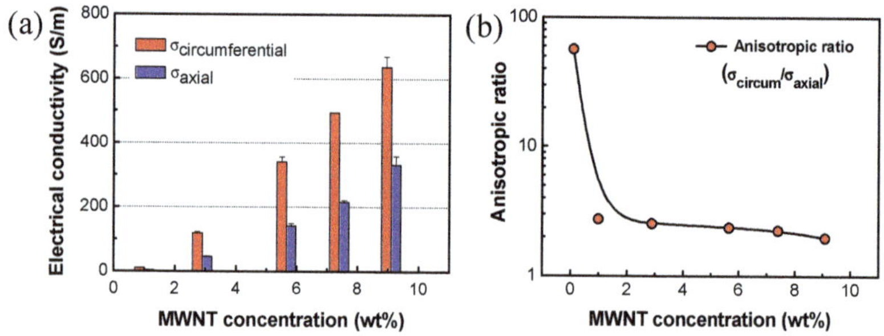

Figure 6. (**a**) Electrical conductivity of MWNT/PDMS composite. (**b**) Anisotropic ratio of electrical conductivity as a function of MWNT contents (axial: axial direction, circum: circumferential direction).

The anisotropy of electrical conductivity resulted from preferential alignment of MWNTs in the circumferential direction [18]. For quantitative analysis of MWNT alignment after the two-roll coating process, polarized Raman spectroscopy was measured and the results are presented in Figure 7. Intensities of polarized Raman spectra are summarized in Table 1.

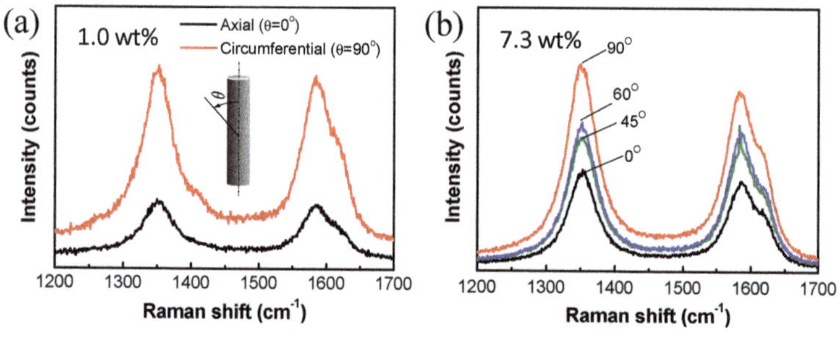

Figure 7. Polarized Raman spectra of MWNT/PDMS composites with MWNT content of (**a**) 1.0 wt%, and (**b**) 7.3 wt%.

Table 1. Polarized Raman intensities of MWNT/PDMS composites by circumferential shearing process.

MWNT Content	Polarization Angle	I_D/I_G	I_D/I_{D0} [1]	I_G/I_{G0} [1]
1.0 wt%	90°	1.05	2.95	3.02
	0°	1.07	-	-
7.3 wt%	90°	1.14	2.04	2.01
	60°	1.07	1.47	1.55
	45°	0.92	1.32	1.63
	0°	1.12	-	-

[1] I_{D0} and I_{G0} represent peak intensities extracted from axial direction ($\theta = 0°$).

The two main typical graphite bands are present in the Raman spectrum of MWNT bundles: The band around 1580 cm^{-1} (G band) assigned to the in-plane vibration of the C–C bond, i.e., sp^2 configuration, and typical of defective graphite-like materials, and the band around 1350 cm^{-1} (D band) activated by the presence of disorder in carbon systems. For the D and G bands, I_D/I_G ratios were in the range of 1.02 to 1.10, regardless of direction and MWNT contents. In this study, little change in the I_D/I_G ratio may be attributed to use of the same quality of CNTs in all the specimens after the dispersion process.

Change in the peak intensity with varying polarization direction was clearly identified in Figure 7 and Table 1. For the composite with MWNT 1 wt% in Figure 7a, the Raman peak intensity from circumferential direction was much larger than that from axial direction. Both I_D and I_G increased up to three times, when polarization direction changed from the axial direction to the circumferential direction, that is, $I_{G,circum}/I_{G,axial}$ and $I_{D,circum}/I_{D,axial}$ are ~3, as shown in Figure 7a. In the same way, intensity ratios for the two directions increased up to two times for the composite with MWNT 7.3 wt%. For the composite in Figure 7b, it is obviously shown that Raman peak intensities are increasing, and both I_D and I_G increased as angles increase from axial (0°) to circumferential (90°) direction, and $I_{G,circum}/I_{G,axial}$ and $I_{D,circum}/I_{D,axial}$ are ~2, which is consistent with the preferential alignment of MWNTs in the circumferential direction during the two-roll coating process, as indicated by the anisotropic electrical conductivity in Figure 6b. The anisotropy of conductivity ($\sigma_{circum}/\sigma_{axial}$) were 2.7 (at 1 wt%) to 2.2 (at 7.3 wt%), which were comparable to $I_{G,circum}/I_{G,axial}$ or $I_{D,circum}/I_{D,axial}$, 3 and 2, respectively, for the corresponding MWNT contents.

Anisotropic physical properties of MWNT composites have been investigated. Ra, et al. [19] fabricated MWNTs-embedded polyacrylonitrile (PAN) nanofiber paper by electrospinning process. Electrical conductivity of the carbonized MWNT/PAN nanofiber papers was highly anisotropic, i.e., the conductivity parallel to the winding direction is about three times higher than that perpendicular to the winding direction. Inoue, et al. [20] fabricated MWNT papers from MWNT webs, and reported high anisotropy ratios of 7.3 in resistivity and of 8.1 in thermal conductivity due to the high alignment of the ultra-long MWNTs having lengths of millimeters. In the polarized Raman spectra, for the G band, the parallel polarization intensity was 4.4 times higher than the perpendicular polarization intensity.

In Figure 7b, peak positions of D and G band maximums were red-shifted when polarization direction moved from axial direction to circumferential direction: 1352.9 to 1348.7 for the D band maximum, and 1586.3 to 1581.2 for the G band maximum. L. Bokobza and J. Zhang [21] reported that weak interaction between MWNTs dispersed in the composite resulted in blue-shift of Raman spectra, compared with the composite with bundled MWNTs having stronger interaction between the conductive nanoparticles. In this study, stronger interaction between MWNTs in the circumferential direction is expected due to the preferential alignment of MWNTs, which may contribute to the redshift of Raman spectra.

3.5. Heating Performance of Printer Fuser

Heating performance of the printer fuser fabricated was demonstrated as presented in Figure 8a. Uniform temperature distribution can be achieved by the optimal dispersion and even alignment of MWNTs, and the uniform thickness of the heating tube. During the heating process, infrared camera images were taken to identify temperature distribution of the printer fuser. The images of temperature distribution, taken at 1.9, 3.8, and 5.2 s, are presented in Figure 8b–d, respectively. The electrical heating was ceased at controller setting temperature of 200 °C, which resulted in maximum tube temperature of 210 °C at 5.8 s, followed by temperature drop with air cooling. Figure 8e is the temperature distribution along the circumferential direction at the positions #1 to #3 in the Figure 8c. For each position, average temperature values with standard deviation in the circumferential direction were 135.7 (±0.6), 142.2 (±0.7), and 138.8 (±0.8) °C, while differences between the maximum and minimum temperatures, (T_{max}-T_{min}), in circumferential direction were 1.7, 2.3, and 3.0 °C, respectively.

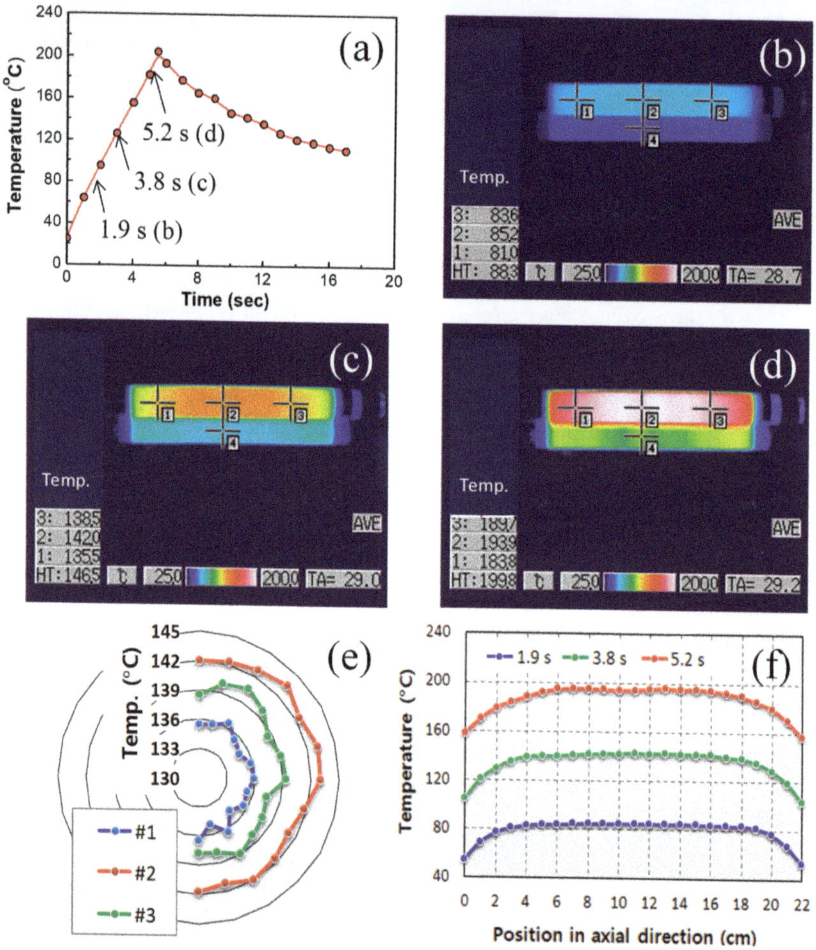

Figure 8. (**a**) Temperature change of MWNT/PDMS fuser with increasing time. Infrared camera images at heating time of (**b**) 1.9 s, (**c**) 3.8 s, and (**d**) 5.2 s. Temperature distribution (**e**) in the circumferential direction at 3.8 s for the three points in Figure 8c, and (**f**) in the axial direction at each moment (1.9 s, 3.8 s, and 5.2 s).

Figure 8f is the temperature distribution in axial direction at the three moments in the Figure 8b–d. The images clearly show that the MWNT/PDMS heating tube possesses the uniform temperature in circumferential and axial directions. Temperature decreases from center to edges in the axial direction.

As the heat transfer was under transient state, generated heats were still conducted to the other parts of the heating and pressing rollers in Figure 8. Therefore, the temperature distribution is obviously enhanced in steady states at a specific operating temperature. The uniform temperature distribution is attributed to uniformity in thickness and electrical conductivity, as analyzed by Equation (3).

Note that the fusing system having MWNT/PDMS composite as heating element reached 180 °C (melting temperature of toner) in 5.0 s and 200 °C in 5.5 s. With this fast heating performance, it can be realized to decrease the first print-out time, which is meaningful for on-demand printing that papers are printed out as soon as customers request printing.

4. Conclusions

A flexible, seamless tube-type heater having uniform temperature distribution and fast thermal response was developed by using a newly-proposed circumferential shearing. Important variables for heating performance, temperature uniformity, and heating speed, were analytically formulated for the composite heating tube. From the theoretical analysis of temperature distribution on the cylindrical composite tube, as shown in Equation (4), the morphological configuration of the heating tube, such as composite thickness and tube diameter, turned out to be one-order more essential for the heating performance than intrinsic material properties such as density, heat capacity, and electrical conductivity. Based on the analytical model, we introduced the circumferential shearing process to fabricate a printer fuser with uniform temperature distribution in the circumferential direction, as the process can provide uniform thickness and shear rate. The circumferential shearing can be a good selection for seamless composite tube structures with outstanding thickness uniformity, especially for the composite paste with high viscosity, which makes difficulty in the fabrication process, such as extrusion.

Supplementary Materials: The following are available online at http://www.mdpi.com/1996-1944/12/20/3283/s1, Figure S1: Optimization process of circumferential shearing with regard to roll speed ratio.

Author Contributions: Conceptualization, S.-E.L. and S.-H.P.; methodology, D.K. and S.-E.L.; software, D.K.; validation, S.-E.L. and S.-H.P.; formal analysis, D.K. and S.-E.L.; investigation, Y.S., S.-H.P. and S.-E.L.; resources, D.K., S.-H.P. and S.-E.L.; data curation, Y.S., D.K., S.-H.P. and S.-E.L.; writing—original draft preparation, Y.S.; writing—review and editing, Y.S., S.-H.P., D.K. and S.-E.L.; visualization, S.-E.L.; supervision, S.-E.L.; project administration, S.-E.L.; funding acquisition, S.-E.L.

Funding: This work was supported by INHA UNIVERSITY Research Grant (60117-1).

Acknowledgments: A special gratitude is cordially expressed to Samsung Advanced Institute of Technology (Suwon-si, Gyeonggi-do, 16678, South Korea) and Taejin (Pyungtaek-si, Gyunggi-do, 17845, South Korea).

Conflicts of Interest: The authors declare no conflict of interest.

References

1. Chou, T.W.; Gao, L.; Thostenson, E.T.; Zhang, Z.; Byun, J.H. An assessment of the science and technology of carbon nanotube-based fibers and composites. *Compos. Sci. Technol.* **2010**, *70*, 1–19.
2. Coleman, J.N.; Khan, U.; Blau, W.J.; Gun'ko, Y.K. Small but strong: A review of the mechanical properties of carbon nanotube–polymer composites. *Carbon* **2006**, *44*, 1624–1652. [CrossRef]
3. Beigbeder, A.; Degee, P.; Conlan, S.L.; Mutton, R.J.; Clare, A.S.; Pettitt, M.E.; Callow, M.E.; Callow, J.A.; Dubois, P. Preparation and characterisation of silicone-based coatings filled with carbon nanotubes and natural sepiolite and their application as marine fouling-release coatings. *Biofouling* **2008**, *24*, 291–302. [CrossRef] [PubMed]
4. Dai, L.; Chang, D.W.; Baek, J.B.; Lu, W. Carbon Nanomaterials for Advanced Energy Conversion and Storage. *Small* **2012**, *8*, 1130–1166. [PubMed]

5. Evanoff, K.; Khan, J.; Balandin, A.A.; Magasinski, A.; Ready, W.J.; Fuller, T.F.; Yushin, G. Towards ultrathick battery electrodes: Aligned carbon nanotube-enabled architecture. *Adv. Mater.* **2012**, *24*, 533–537. [CrossRef] [PubMed]
6. Yoon, Y.H.; Song, J.W.; Kim, D.; Kim, J.; Park, J.K.; Oh, S.-K.; Han, C.-S. Transparent film heater using single-walled carbon nanotubes. *Adv. Mater.* **2007**, *19*, 4284–4287.
7. Jeong, Y.G.; Jeon, G.W. Microstructure and performance of multiwalled carbon nanotube/m-aramid composite films as electric heating elements. *ACS Appl. Mater. Interf.* **2013**, *5*, 6527–6534.
8. Yu, C.; Choi, K.; Yin, Y.; Grunlan, J.C. Light-weight flexible carbon nanotube based organic composites with large thermoelectric power factors. *ACS Nano* **2011**, *5*, 7885–7892. [CrossRef] [PubMed]
9. Park, S.H.; Cho, E.H.; Sohn, J.; Theilmann, P.; Chu, K.; Lee, S.; Sohn, Y.; Kim, D.; Kim, B. Design of multi-functional dual hole patterned carbon nanotube composites with superhydrophobicity and durability. *Nano Res.* **2013**, *6*, 389–398. [CrossRef]
10. Fujii, M.; Zhang, X.; Xie, H.; Ago, H.; Takahashi, K.; Ikuta, T.; Abe, H.; Shimizu, T. Measuring the thermal conductivity of a single carbon nanotube. *Phys. Rev. Lett.* **2005**, *95*, 065502. [CrossRef] [PubMed]
11. Berber, S.; Kwon, Y.K.; Tománek, D. Unusually high thermal conductivity of carbon nanotubes. *Phys. Rev. Lett.* **2000**, *84*, 4613–4616. [CrossRef] [PubMed]
12. Ryu, Y.; Yin, L.; Yu, C. Dramatic electrical conductivity improvement of carbon nanotube networks by simultaneous de-bundling and hole-doping with chlorosulfonic acid. *J. Mater. Chem.* **2012**, *22*, 6959–6964. [CrossRef]
13. Lee, S.-E.; Sohn, Y.; Chu, K.; Kim, D.; Park, S.-H.; Bae, M.; Kim, D.; Kim, Y.; Han, I.T.; Kim, H.-J. Suppression of negative temperature coefficient of resistance of multiwalled nanotube/silicone rubber composite through segregated conductive network and its application to laser-printing fusing element. *Org. Electron.* **2016**, *37*, 371–378. [CrossRef]
14. Yan, J.; Jeong, Y.G. Multiwalled carbon nanotube/polydimethylsiloxane composite films as high performance flexible electric heating elements. *Appl. Phys. Lett.* **2014**, *105*, 051907. [CrossRef]
15. Fang, H.; Zhao, Y.; Zhang, Y.; Ren, Y.; Bai, S.-L. Three-dimensional graphene foam-filled elastomer composites with high thermal and mechanical properties. *ACS Appl. Mater. Interf.* **2017**, *9*, 26447–26459. [CrossRef] [PubMed]
16. Lee, S.E.; Cho, S.J.; Kim, H.J.; Han, I.T.; Sohn, Y.C. Advanced catalyst design induced enhancement of multi-walled nanotube debundling and electrical conductivity of multi-walled nanotube/silicone composites. *RSC Adv.* **2016**, *6*, 48120–48128. [CrossRef]
17. Pradhan, N.R.; Duan, H.; Liang, J.; Iannacchione, G.S. The specific heat and effective thermal conductivity of composites containing single-wall and multi-wall carbon nanotubes. *Nanotechnol.* **2009**, *20*, 245705. [CrossRef] [PubMed]
18. Bauhofer, W.; Kovacs, J.Z. A review and analysis of electrical percolation in carbon nanotube polymer composites. *Compos. Sci. Technol.* **2009**, *69*, 1486–1498. [CrossRef]
19. Ra, E.J.; An, K.H.; Kim, K.K.; Jeong, S.Y.; Lee, Y.H. Anisotropic electrical conductivity of MWCNT/PAN nanofiber paper. *Chem. Phys. Lett.* **2005**, *413*, 188–193. [CrossRef]
20. Inoue, Y.; Suzuki, Y.; Minam, Y.; Muramatsu, J.; Shimamura, Y.; Suzuki, K.; Ghemes, A.; Okada, M.; Sakakibara, S.; Mimura, H.; et al. Anisotropic carbon nanotube papers fabricated from multiwalled carbon nanotube webs. *Carbon* **2011**, *49*, 2437–2443. [CrossRef]
21. Bokobza, L.; Zhang, J. Raman spectroscopic characterization of multiwall carbon nanotubes and of composites. *Exp. Poly. Lett.* **2012**, *6*, 601–608. [CrossRef]

materials

Article

Surface Roughening of Electrolyte Membrane for Pt- and Ru-Sputtered Passive Direct Methanol Fuel Cells

Wonyeop Jeong [1,†], Gu Young Cho [2,†], Suk Won Cha [1,*] and Taehyun Park [3,*]

1 School of Mechanical and Aerospace Engineering, Seoul National University, 1 Gwanak-ro, Gwanak-gu, Seoul 08826, Korea; starryshot@snu.ac.kr
2 Department of Mechanical Engineering, Dankook University, 152, Jukjeon-ro, Suji-gu, Yongin-si, Gyeonggi-do 16890, Korea; guyoungcho@dankook.ac.kr
3 School of Mechanical Engineering, Soongsil University, 369 Sangdo-ro, Dongjak-gu, Seoul 06978, Korea
* Correspondence: swcha@snu.ac.kr (S.W.C.); taehyunpark@ssu.ac.kr (T.P.); Tel.: +82-2-880-1700 (S.W.C.); +82-2-820-0669 (T.P.)
† These authors contributed equally to this work.

Received: 16 October 2019; Accepted: 28 November 2019; Published: 29 November 2019

Abstract: Platinum (Pt) and ruthenium (Ru) were sputtered on an electrolyte membrane and it was used as a membrane-electrode assembly for passive direct methanol fuel cells (DMFCs) operating with high concentration methanol solution (4 M). Thick (Pt of 300 nm and Ru of 150 nm) and thin (Pt of 150 nm and Ru of 75 nm) sputtered catalysts were prepared and their performance was first evaluated to find out the best sputtering conditions showing higher performance. Subsequently, four electrolyte membranes with different surface roughness were prepared to investigate its influence on the performance. As a result, the performance of the passive DMFC showed increasing tendency as the roughness is low, while the performance was decreased as the roughness was high, indicating there exists an optimal roughness of the electrolyte membrane. It was further investigated through morphological study through electron microscopy that such performance variation is attributed to the surface of sputtered Pt–Ru catalyst on the rough electrolyte membrane that adequate roughness induces the increase of reactive area while a too rough surface bears the poor contact of it with gas-diffusion layer.

Keywords: direct methanol fuel cells; sputter; sandpaper; roughness; electrochemical impedance spectroscopy; polarization

1. Introduction

Many researchers have been working on the investigation of the alternatives of current market-leading lithium-ion batteries to store more energy in a limited volume. One of the technologies is considered as a fuel cell because fuel cells have potentially higher energy density than lithium-ion batteries [1–3]. In addition, fuel cells have other advantageous characteristics such as scalability, eco-friendliness, high efficiency, and no need to secure recharging time [4–7]. That is why many portable fuel cell prototypes are actively coming out to markets for the purpose of replacing batteries [8–10].

Among various fuel cell types, polymer electrolyte membrane fuel cells (PEMFCs) are the most famous type because they operate at low temperature (<100 °C) and their technological level is close to commercialization level. Fuel cell electric vehicles, drones including unmanned aerial vehicles and quadrotors, and power plants already manufactured or installed prove it [11–14]. Here, it is noticeable that the recent world-champion record of the flight time of quad-/hexa-rotors exceeds 12 h and it has been enabled by employing PEMFCs with liquid hydrogen storage [15]. Comparing it with the fact that the top-flight time by any batteries is shorter than 30 min, the fuel cells and their features they could enable are amazing.

The use of hydrogen as a fuel, however, bears a problem: The hydrogen storage technology is not sufficiently matured than fuel cells [16–18]. In addition, hydrogen storage is directly related to safety, so it makes the PEMFCs still distant from complete commercialization. In order to overcome this, direct methanol fuel cells (DMFCs) have been investigated for decades. The basic theory and the structure of operation are the same as PEMFCs, but they are distinguished in that a methanol solution instead of hydrogen is supplied and ruthenium (Ru) as well as platinum (Pt) is normally added in an electro-catalyst. Here, if the chemical energy resource is stored as a methanol, it is enabled theoretically that the DMFC systems can store more energy than normal PEMFC systems in a same volume.

The problems of the DMFCs are that they require more novel metals (Pt and Ru) than PEMFCs, the resulting electrochemical performance is still lower than PEMFCs, and their system is highly complicated. The third problem comes from the feature that the water management and resulting methanol concentration within the system is highly important to maximize the performance and secure the long-term durability. One way to resolve this problem is to simplify the fuel supply system in an anode and cathode by designing them "fully passive". That is, the supply of whole reactants in DMFC fully depends on free convection. Many reports about this fully passive DMFCs are already in literature [2,19–27].

In this study, we especially employed the sputtering method to fabricate the membrane-electrode assembly (MEA) for passive DMFCs. It is because two advantages could be achieved from this approach: Lowering the use of novel metals and drive down the manufacturing cost of DMFCs. It is because the sputtering method has been experimentally proved to reduce the use of novel metals. Although most of sputtering-based fuel cells are about PEMFCs [28–32], we expected that the application of this technique to DMFCs would result in the same effect as PEMFCs. In addition, industrial infrastructure of sputtering is already tremendous due to the development of semiconductors industry. It means that we could maximize the advantages of the development of this sputter-based MEA fabrication approach for passive DMFCs. Unfortunately, to our knowledge, no report about this sputtered catalyst-based passive DMFCs can be found in literature. With this sputtering approach, this study also investigates the effect of roughness of the electrolyte membrane on the performance of the passive DMFCs. The MEAs with four roughnesses were prepared and their micro-morphologies are investigated through scanning-electron microscopy (SEM). Moreover, the thickness of the sputtered Pt and Ru is tested to find out the structural effect of sputtered layers (Note too thin catalyst would not activate the oxidation of the methanol while too thick could disturb the diffusion of reactants.). Finally, the performances of the passive DMFCs with four MEAs are measured and compared to find out the optimal roughness showing the best electrochemical performance and the relation between the roughness of an electrolyte membrane and the resulting performance.

2. Materials and Methods

Four types of MEAs were fabricated to compare the effects of surface roughness of an electrolyte membrane on electrochemical characteristics of passive DMFCs, as shown in Figure 1a. First, the standard MEA with a pristine electrolyte membrane (Nafion® 117, DuPont Co., Midland, MI, USA) was fabricated using sputter. Pt and Ru were deposited sequentially on the anode side of Nafion® 117 to deposit bi-layered catalyst. 100 and 200 W of DC sputtering power were applied to Pt and Ru target, respectively. The sputtering conditions were 12.0 Pa of Ar gas pressure and room temperature of substrate in all cases. Two bi-layered catalysts were fabricated first: Thicknesses of Pt–Ru catalysts of 300 and 150 nm, and 150 and 75 nm, respectively. These two samples were characterized and the thickness showing the best performance was selected. After depositing bi-layered Pt–Ru catalysts, cathodic catalyst layers were deposited on the other side of the MEA by sputtering again. Pt was deposited with 100 W of DC power and 12.0 Pa of Ar pressure. The electrochemically reactive area was precisely controlled and defined by using the physical masking tape with a hole of 1×1 cm^2. In the fabrication of the MEAs with rough electrolyte surfaces, Nafion® 117 membrane was rubbed with sandpapers to change the surface roughness. Three kinds of sandpapers (600, 2000, and 4000 grit,

Daesung Abrasive Co., Yonki, Chungnam, Korea) with different roughness were used to vary the surface roughness. When the membrane is roughened, its surface becomes opaque. We scrubbed the center of the membrane until it became uniformly opaque over a wider range than the catalysts deposition area. After preparing the membrane with three different roughnesses, Pt–Ru catalysts were deposited using the same sputtering conditions as described above. Digital camera images of the as-prepared four MEAs are indicated in Figure 1b.

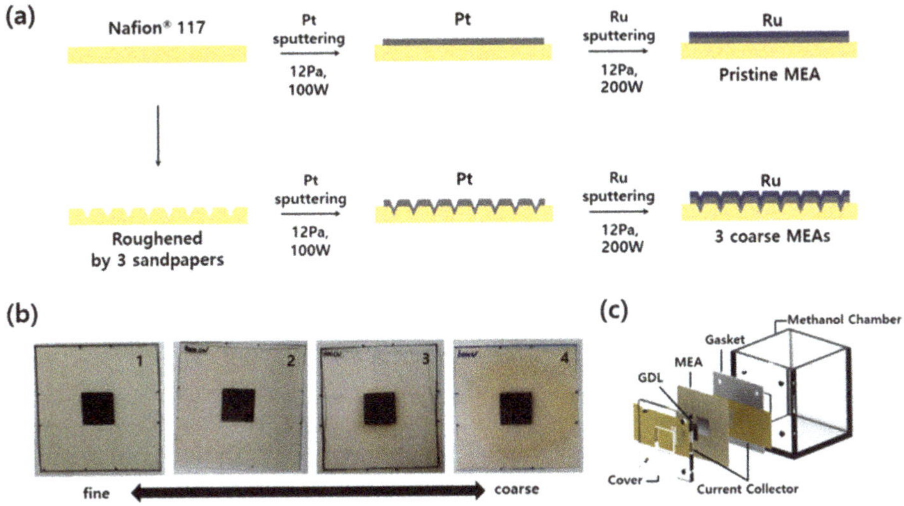

Figure 1. (**a**) Schematic of the fabrications of the membrane-electrode assembly (MEAs) with different roughnesses. (**b**) Real images of as-prepared MEAs. (**c**) Exploded image of the fully passive direct methanol fuel cells (DMFC) used in this study. This is a figure, Schemes follow the same formatting.

After preparing MEAs, the four-step pretreatment process was performed to secure the high protonic conductivity of the electrolyte membrane. This process is intended to remove impurities and recover the sulfonic acid group in the Nafion® 117 membrane. The procedure was as follows: (1) Boiling for 1 h in 5 vol% H_2O_2 solution, (2) boiling for 1 h in deionized water, (3) boiling for 1 h in 0.5 M H_2SO_4 solution, and (4) boiling for 1 h in deionized water [33].

Electrochemical characterizations of the as-fabricated MEAs and the passive DMFCs comprising thereof were conducted using the custom-made passive DMFC setup, as shown in Figure 1c. It consists of a methanol chamber, two polytetrafluoroethylene gaskets, two Au-coated stainless-steel current collectors, two gas-diffusion layers (GDLs, Sigracet 39BC, SGL Carbon Co., Wiesbaden, Germany) with micro-porous layers at one side, and one endplate. All components were assembled tightly by four bolts and nuts. The 0.1 mm thick current collectors were fabricated by laser-cutting and Au was subsequently sputtered on one side of it. For Au deposition, 200 W of DC power and 0.67 Pa of Ar pressure at room temperature were used. The thickness of the as-deposited Au layer was 20 nm. The volume of methanol chamber was 5 × 5 × 5 cm^3 and a 4 M methanol solution was supplied as a fuel into the chamber.

The surface and cross-sectional images of the as-prepared MEAs were obtained and investigated using a field-emission scanning-electron microscope (FE-SEM, Zeiss Supra 55VP, Carl Zeiss, Oberkochen, Germany) and a focused-ion beam SEM (FIB-SEM, Nova 600, FEI Company, Hilsboro, OR, USA). In the FIB process, a platinum buffer layer was deposited to protect the electrode layer before milling. The main trench was milled with an ion current of 1 nA and a lower current of 0.5 nA was used to polish the cross-section for imaging. The performance measurement was carried out at room temperature using a potentiostat (Reference 600, Gamry Instruments Inc., Warminster, PA, USA). Electrodes of

MEAs were simply exposed to the 4 M methanol and ambient air, respectively. Thus, fuel and air were supplied to each electrode by free convection, not forced nor controlled. Polarization characteristics were investigated by measuring current–voltage (I–V) curves. The measurement started from the open-circuit voltage (OCV) of the fuel cell, and swept towards 0.1 V. The voltage sweep speed was 0.01 V/s, and resulting current was monitored. It was halted at 0.1 V compared to RHE (reversible hydrogen electrode). Electrochemical impedance spectra (EIS) were measured and visualized on Nyquist plot. A sinusoidal voltage input with an amplitude of 0.01 V and a frequency range from 10^6 to 0.1 Hz at 0.1 V compared to RHE was applied to the fuel cell and the resulting current response was measured.

3. Results and Discussion

The effect of the thickness of an anodic catalyst layer on the electrochemical performance of the passive was first investigated, as shown in Figure 2a, b. It is because too thin catalyst cannot activate the polarization while too thick could disturb the diffusion of the methanol and air. Same phenomena can be found elsewhere that the PEMFCs should be clamped with appropriate pressure because too strong clamping pressure could stuff the pores inside GDLs, thereby resulting in more frequent flooding and high activation overvoltage by the lowered concentration of the reactants near reactive sites. On the other hand, weak clamping force could bear the high contact resistance between GDLs and bipolar plates [34]. It is speculated that this can be compared by the experimental results in Figure 2 because the phenomena and resulting performance variation resemble: Too thick catalyst could disturb the diffusion of reactants. In real, as shown in Figure 2a, the DMFC with a thin catalyst (75 nm thick Ru on 150 nm thick Pt) shows higher peak power density (0.53 mW/cm^2) than that with thick (0.33 mW/cm^2) one. Although both power curves are apparently in increasing tendency so we could not mention a "peak" power density, it seems the DMFC with thicker catalyst could not outperform the thin one at high current density region. It is thought that such difference is attributed to the disturbance of mass transport by the thick catalyst. According to the EIS results indicated in Figure 2b, it is no doubt that charge transport resistance is a dominant factor of the final electrochemical performance because in both cases, the charge transport resistance (122–275 $\Omega\cdot$cm^2) is extremely higher than the ohmic resistance (<5 $\Omega\cdot$cm^2). In addition, the charge transfer resistance of the DMFC with thick catalyst is higher than that with thin catalyst. If considering the activation overvoltage is described by the Butler–Volmer behavior, increasing charge transfer resistance with the increasing Pt cannot be explained. In this case, it is speculated that the stuffed pores by thicker layer inactivated the catalyst. It corresponds with the observation in Figure 2a that the OCV of the DMFC with thin catalyst is lower (0.177 V) than that with thick catalyst (0.216 V). The OCV can be affected by two factors: Concentration of reactants and electrical insulation [1]. It is then thought that, as expected from the difference of charge transfer resistances, such pore-stuffing effect also blocked the methanol crossover, thereby finally increasing the OCV. Interestingly, in spite of the high starting voltage for thicker electrode case, the final performance is turned around as the current density increases, meaning the charge transfer resistance is a dominant factor as mentioned.

The performance difference between thick and thin catalyst layers could also be confirmed by surface morphology in Figure 3. In the case of the thick catalyst layer, as shown in Figure 3a, cracks and delamination of catalysts were seen. Such cracks and delamination came from the expansion coefficient difference between the membrane and catalysts layer. Nafion® 117 membranes must be activated prior to characterizations. However, during activation process, Nafion® membrane absorbed large amounts of water and expanded in volume. At that time, defects occurred in the catalyst layer and the thicker the catalyst layer is, the more severely affected by the volume change [35]. When the catalysts were deposited on the membrane by sputtering, the triple phase boundaries (TPBs) are formed only at the interface between the electrolyte and the electrode. Therefore, due to such cracks and delamination, the thick catalysts layer MEA had less TPBs than the thin one, and as a result, the charge transfer resistance was increased, as shown in Figure 2b.

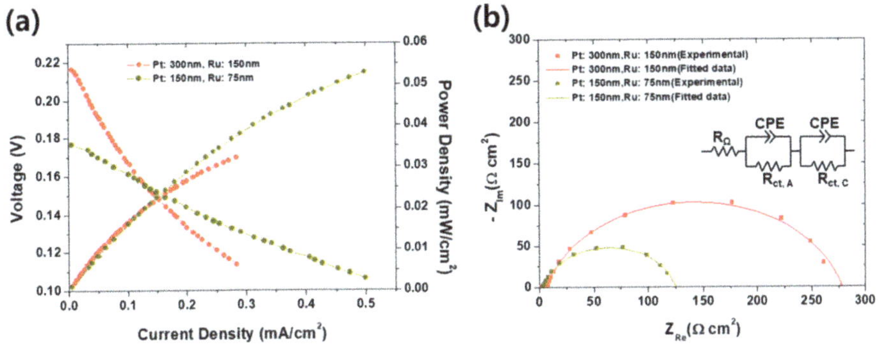

Figure 2. (**a**) Current–voltage and current–power curves, and (**b**) electrochemical impedance spectra (EIS) measured at 0.1 V compared to reversible hydrogen electrode (RHE) corresponding to (**a**) of the passive DMFCs with anodic catalysts of 150 nm thick Ru on 300 nm thick Pt and 75 nm thick Ru on 150 nm thick Pt.

Figure 3. FE-SEM images of the surfaces with different catalysts thickness. (**a**) 150 nm thick Ru on 300 thick Pt, and (**b**) 75 nm thick Ru on 150 nm thick Pt.

Although not thoroughly investigated, since 150 nm thick Pt and 75 nm thick Ru was found to give a higher electrochemical performance in Figure 2, such sputtered catalyst was applied similarly to roughened Nafion® 117. Figure 4 presents images of FIB images of the bi-layered Pt–Ru catalysts on anode side of Nafion® 117. As shown in Figure 4, all MEAs have very similar thickness of bi-layered Pt–Ru catalysts (150 nm thick Pt and 75 nm thick Ru). It means that catalyst-coated layers (CCLs) were successfully prepared on Nafion® 117 MEAs by sequential sputtering process without any complicated solution-based spray processes. In addition, it is clearly observable that the roughness of CCLs are successfully varied in Figure 4. The maximum height roughness (R_{max}) for pristine, 4000, 2000, and 600 grit roughened CCLs are 82, 100, 120, and 250 nm, respectively. Here, the surface of the MEA which is marked with the white dashed line in each image in Figure 4 becomes rougher as the roughness of sandpapers (4000, 2000, and 600 grit) which were used in pre-treatment becomes rougher from Figure 4a–d. Therefore, it is confirmed that the interface length between Pt catalysts and Nafion® 117 was increased as the roughness of MEA was increased.

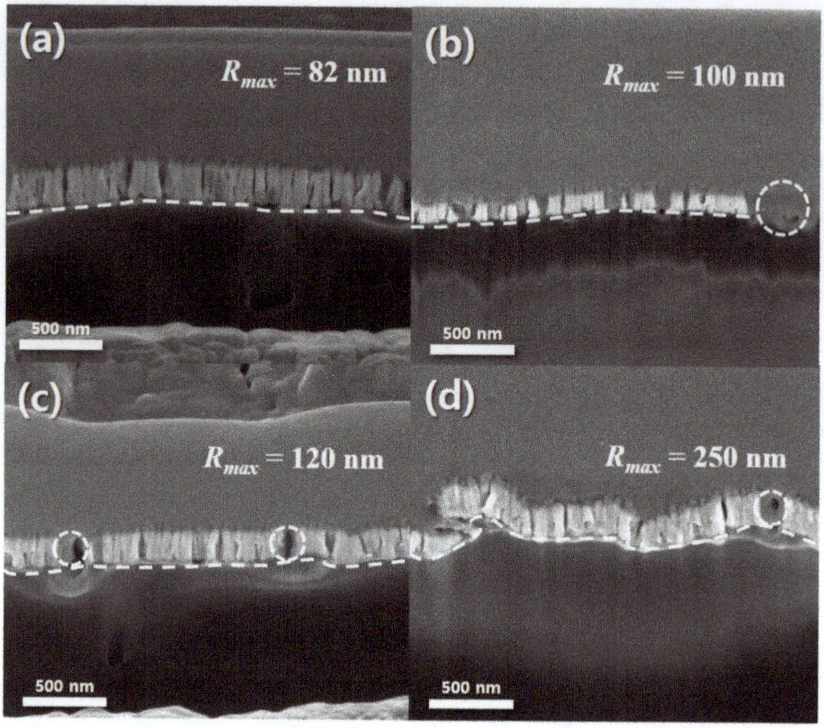

Figure 4. FIB-SEM images of the sputtered Pt and Ru on (**a**) pristine membrane, and roughened membranes by (**b**) 4000, (**c**) 2000, and (**d**) 600 grit sand papers. The measured R_{max} for four surfaces are 82, 100, 120, and 250 nm for (**a**), (**b**), (**c**), and (**d**), respectively.

The surface morphologies of the bi-layered Pt–Ru catalysts on Nafion® 117 were investigated further in order to enunciate the relation between the roughness and the electrochemical performance, as shown in Figure 5. All MEAs show rough and porous surfaces in spite of the deposited 250 nm thick sputtered Pt–Ru catalysts. Especially, it seems that the porosity of the surface of a MEA is slightly enlarged as the surface roughness increases. Here, it is noted that the surface morphology of the thin film fabricated by sputtering process is strongly dependent on both the surface roughness of substrate and deposition conditions of sputtering [36–39]. When considering that the deposition process of bi-layered Pt–Ru catalysts is identical for all MEAs, such differences of the porosities is mainly attributed to the roughness of the MEAs. It is also noticed that the size of pores is <1 μm, meaning that such morphologies would not disturb the diffusion of any reactants (Air and methanol) of the DMFC.

Finally, the electrochemical characteristics of the passive DMFCs with various roughness MEAs were investigated, as shown in Figure 6. In these evaluations, we used 4 M methanol solution to pursue the simplification of the DMFC systems for portable applications as mentioned above [9]. Here, the grit numbers (4000, 2000, and 600) are the levels of surface roughness of sandpapers which are used in pretreatment of MEAs before the sputter process. According to the Nernst equation, the theoretical OCV of a DMFC is 1.199 V compared to RHE [1]. However, the OCVs indicated in Figure 6 are significantly low regardless of surface roughness of MEAs, which is same as the result of Figure 2a. The OCVs are 0.18, 0.23, 0.23, and 0.18 V for a pristine, 4000, 2000, and 600 grit roughened MEA, respectively. Here again, the low OCVs of DMFCs can be explained by the methanol crossover in electrolyte membranes, as seen in Figure 2a [21]. Methanol crossover from anode to cathode causes the methanol oxidation reactions at the cathode side of the fuel cell. It would end up lowering the OCVs of DMFCs significantly. In this study, in order to prevent the drop of OCV, Nafion® 117 was used as an electrolyte membrane to reduce the methanol crossover through electrolyte because of its sufficient thickness than other electrolyte membrane generally used for PEMFCs (178 μm thick for Nafion® 117 compared to <50 μm thick for normal PEMFCs) [40]. Here, the custom made DMFC of this study shows no discernible defects (no leakage of methanol solution from a chamber). Therefore, we could first conclude that these considerably low OCVs were caused mainly by two reasons: Lack of electrochemical catalyst and insufficient TPBs at the interface between Nafion® 117 and bi-layered Pt–Ru catalysts, which are also the case in Figure 2a. Compared to the general Pt/C- or Pt–Ru/C-sprayed MEAs, 150 nm thick sputtered Pt and 75 nm thick sputtered Ru are extremely lower than the MEAs fabricated by spraying [41]. Theoretical OCV means that the difference of electrical potential at open circuit is measured without any ohmic losses generated from electrical resistances. In real cases, however, enough current is required to measure OCVs of fuel cells to overcome contact resistances between components of evaluation system and the resistances of external wires. In order to generate enough current from the fuel cell, both sufficient anodic and cathode electrochemical reactions should be accompanied with. In other studies, all MEAs have the loading of Pt in normal range ($\geq$1.0 mg_{Pt}/cm^2) [42–45]. However, in this study, if the Pt layer was 1 cm^2, 150 nm thick, and super-dense (no pores), Pt loading would be 0.32 mg/cm^2. In addition, Pt/C catalyst, which was often used as catalyst for DMFCs, was generally applied together with Nafion® solution to make mixed ionic-electronic conductor (MIEC) to increase TPBs. However, catalysts sputtered MEAs only had TPBs at the interface between the electrolyte and the electrode. Thus, it is thought that the significantly low OCVs in Figure 6 is due to insufficient loading of Pt and deficient TPBs.

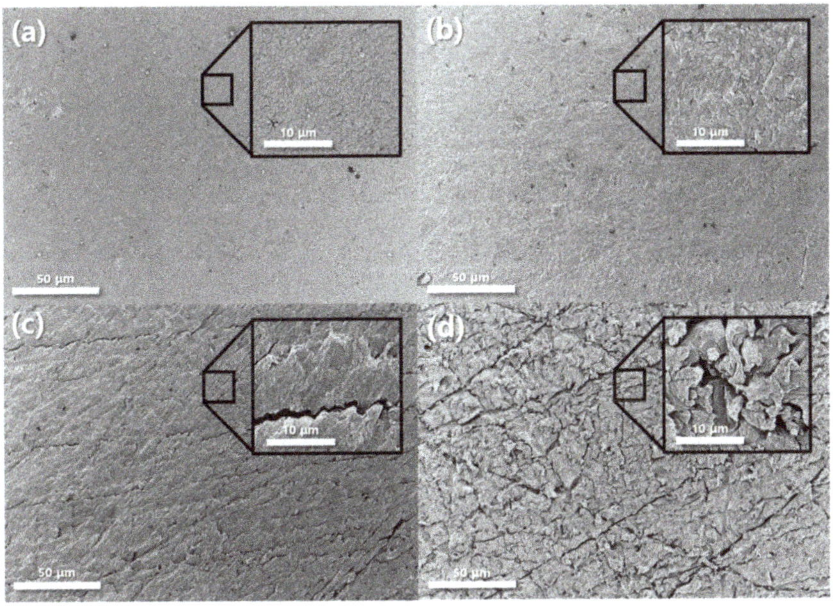

Figure 5. FE-SEM images of the surfaces of sputtered Pt–Ru on (**a**) pristine, (**b**) 4000, (**c**) 2000, and (**d**) 600 grit roughened Nafion® 117 membranes. The insets in each image indicate the magnified surface morphology.

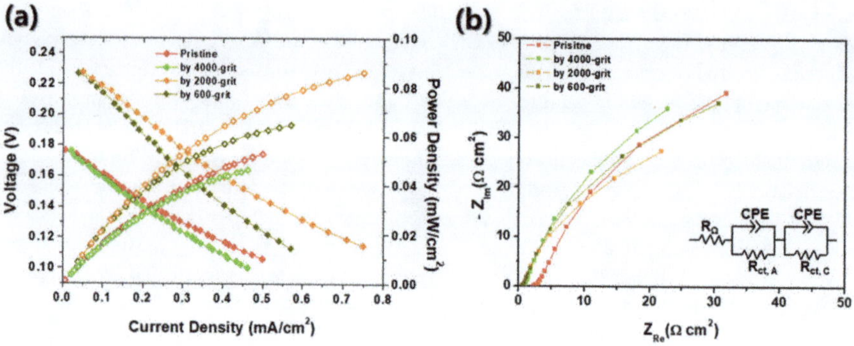

Figure 6. (**a**) Current–voltage and current–power curves and (**b**) EIS spectra of passive DMFCs with four different MEAs: Pristine, roughened by 4000, 2000, and 600 grit.

Interestingly, there are differences between performances of DMFCs with four different MEAs. According to Figure 6a, the DMFC with 2000 grit rubbed MEA has the highest performance among all samples, which is an OCV of 0.23 V and power density of 0.086 mW/cm^2. Other fuel cells have relatively lower performances, which are 0.053 mW/cm^2 for a pristine MEA, 0.065 mW/cm^2 for a 600 grit rubbed MEA, and 0.046 mW/cm^2 for a 4000 grit rubbed MEA. These results show the strong dependence between surface roughness of MEAs and performance of fuel cells. Accordingly, it can be confirmed that Figure 6a depicts the relations between surface roughness and performances of fuel cells. As shown in Figure 6a, the peak power density and OCV increase as the roughness of an electrolyte membrane increases at low roughness range. However, excessive roughness of MEA causes the reduction of performance and OCVs. It is speculated that the inordinate surface roughness of the electrolyte membrane could bring about the electrical disconnection between sputtered catalyst themselves (Figures 4 and 5). Or the contact area between the sputtered catalyst layer and GDL could be deteriorated due to the highly roughened architecture of the electrolyte membrane. According to the EIS results indicated in Figure 6b, the charge transport resistance is a dominant factor in each case. Although the charge transport resistance does not appear as a semicircle on the Nyquist plot, it can be seen that it is extremely larger than ohmic resistance (<2.2 Ω·cm^2). Additionally, the size of charge transport resistance can be compared for each MEA, and it can be seen that it has the lowest charge transport resistance at 2000 grit MEA, which has the best performance. It was thought that the roughened MEA had increased the TPBs, thereby decreasing the charge transport resistance.

Further investigations about the relation between the roughness of the electrolyte membrane and the electrochemical performance are required to find out the deep science inside the passive DMFCs and their sputtered catalysts. It could be the optimization of the structures of sputtered layers, thicknesses, and applications of proper GDLs. However, the findings in this study suggest two meaningful results: Sputtered Pt and Ru can be used as an electro-catalyst for passive DMFCs and their structure (thickness of Pt/Ru and roughness of an electrolyte membrane) is a significant parameter affecting the performance of the DMFCs.

4. Conclusions

In this study, Pt and Ru were sputtered directly on an electrolyte membrane (Nafion® 117) and it was applied as an MEA for passive DMFCs for the first time. Especially, the influence of the surface roughness of an electrolyte membrane on electrochemical characteristics of passive DMFCs was investigated. In addition, it was operated by using a methanol solution of very high concentration (4 M). As a result, bi-layered Pt–Ru catalysts layer was successfully fabricated by sputtering and surface roughness of an electrolyte membrane was precisely controlled using sandpapers with different roughness level, which were confirmed by SEM images. The performance of the passive DMFCs with the as-prepared MEAs were improved as the roughness of MEA was increased. However, excessive roughness induced the deterioration of the performance. The passive DMFC with the optimal roughness of the electrolyte membrane improved the OCV by 22% and enhanced performance compared to a fuel cell with a pristine MEA by 38%. Although further investigation such as the optimization of the structure and thickness of Pt and Ru, finding proper GDLs, and so forth is required to further improve this type of MEAs and DMFCs, we believe that results of this study can contribute to the reduction of novel metals (Pt and Ru) by using sputtering process and resulting increased possibility of the commercialization of the passive DMFCs.

Author Contributions: Conceptualization, T.P.; Methodology, G.Y.C. and T.P.; Validation, W.J. and G.Y.C.; Formal analysis, W.J., G.Y.C., and T.P.; Investigation, W.J., G.Y.C., and T.P.; Resources, T.P.; Data curation, W.J. and G.Y.C.; Writing—original draft preparation, G.Y.C. and T.P.; Writing—review and editing, W.J. and T.P.; Supervision, S.W.C. and T.P.; Project administration, S.W.C. and T.P.; Funding acquisition, S.W.C. and T.P.

Funding: This research was funded by an NRF grant funded by the Ministry of Science and ICT, Republic of Korea (NRF-2019R1A4A1025848).

Conflicts of Interest: The authors declare no conflict of interest.

References

1. O'Hayre, R.; Cha, S.-W.; Colella, W.; Prinz, F.B. *Fuel Cell Fundamentals*, 3rd ed.; Wiley: Hoboken, NJ, USA, 2016; ISBN 978-1119113805.
2. Faghri, A.; Guo, Z. An innovative passive DMFC technology. *Appl. Therm. Eng.* **2008**, *28*, 1614–1622. [CrossRef]
3. Dyer, C.K. Fuel cells for portable applications. *J. Power Source* **2002**, *106*, 31–34. [CrossRef]
4. Mahato, N.; Banerjee, A.; Gupta, A.; Omar, S.; Balani, K. Progress in material selection for solid oxide fuel cell technology: A review. *Prog. Mater. Sci.* **2015**, *72*, 141–337. [CrossRef]
5. Ormerod, R.M. Solid oxide fuel cells. *Chem. Soc. Rev.* **2003**, *32*, 17–28. [CrossRef] [PubMed]
6. Badwal, S.P.S.; Foger, K. Solid oxide electrolyte fuel cell review. *Ceram. Int.* **1996**, *22*, 257–265. [CrossRef]
7. Stambouli, A.B.; Traversa, E. Solid oxide fuel cells (SOFCs): A review of an environmentally clean and efficient source of energy. *Renew. Sustain. Energy Rev.* **2002**, *6*, 433–455. [CrossRef]
8. Wilberforce, T.; Alaswad, A.; Palumbo, A.; Dassisti, M.; Olabi, A.G. Advances in stationary and portable fuel cell applications. *Int. J. Hydrogen Energy* **2016**, *41*, 16509–16522. [CrossRef]
9. Kamarudin, S.K.; Achmad, F.; Daud, W.R.W. Overview on the application of direct methanol fuel cell (DMFC) for portable electronic devices. *Int. J. Hydrogen Energy* **2009**, *34*, 6902–6916. [CrossRef]
10. Oszcipok, M.; Zedda, M.; Hesselmann, J.; Huppmann, M.; Wodrich, M.; Junghardt, M.; Hebling, C. Portable proton exchange membrane fuel-cell systems for outdoor applications. *J. Power Sources* **2006**, *157*, 666–673. [CrossRef]
11. Elmer, T.; Worall, M.; Wu, S.; Riffat, S.B. Fuel cell technology for domestic built environment applications: State of-the-art review. *Renew. Sustain. Energy Rev.* **2015**, *42*, 913–931. [CrossRef]
12. Wee, J.H. Applications of proton exchange membrane fuel cell systems. *Renew. Sustain. Energy Rev.* **2007**, *11*, 1720–1738. [CrossRef]
13. Oh, T.H. Conceptual design of small unmanned aerial vehicle with proton exchange membrane fuel cell system for long endurance mission. *Energy Convers. Manag.* **2018**, *176*, 349–356. [CrossRef]
14. Kim, T. NaBH4 (sodium borohydride) hydrogen generator with a volume-exchange fuel tank for small unmanned aerial vehicles powered by a PEM (proton exchange membrane) fuel cell. *Energy* **2014**, *69*, 721–727. [CrossRef]
15. Robodex and Intelligent Energy to showcase drone fuel cell power modules at Fukushima Robot Test Field in Japan. Available online: https://www.intelligent-energy.com/news-and-events/company-news/2019/07/10/robodex-and-intelligent-energy-to-showcase-drone-fuel-cell-power-modules-at-fukushima-robot-test-field-in-japan/ (accessed on 8 November 2019).
16. Broom, D.P.; Webb, C.J.; Hurst, K.E.; Parilla, P.A.; Gennett, T.; Brown, C.M.; Zacharia, R.; Tylianakis, E.; Klontzas, E.; Froudakis, G.E.; et al. Outlook and challenges for hydrogen storage in nanoporous materials. *Appl. Phys. A Mater. Sci. Process.* **2016**, *122*, 151. [CrossRef]

17. Yu, X.; Tang, Z.; Sun, D.; Ouyang, L.; Zhu, M. Recent advances and remaining challenges of nanostructured materials for hydrogen storage applications. *Prog. Mater. Sci.* **2017**, *88*, 1–48. [CrossRef]
18. Suri, S.U.K.; Siddique, M. Novel and Optimized Techniques for Storage and Transportation of Hydrogen: Perspectives and Challenges. *J. Appl. Emerg. Sci.* **2019**, *9*, 8–15.
19. Mallick, R.K.; Thombre, S.B. Performance of passive DMFC with expanded metal mesh current collectors. *Electrochim. Acta* **2017**, *243*, 299–309. [CrossRef]
20. Chen, R.; Zhao, T.S. A novel electrode architecture for passive direct methanol fuel cells. *Electrochem. Commun.* **2007**, *9*, 718–724. [CrossRef]
21. Kho, B.K.; Bae, B.; Scibioh, M.A.; Lee, J.; Ha, H.Y. On the consequences of methanol crossover in passive air-breathing direct methanol fuel cells. *J. Power Sources* **2005**, *142*, 50–55. [CrossRef]
22. Mallick, R.K.; Thombre, S.B.; Shrivastava, N.K. A critical review of the current collector for passive direct methanol fuel cells. *J. Power Sources* **2015**, *285*, 510–529. [CrossRef]
23. Bae, B.; Kho, B.K.; Lim, T.H.; Oh, I.H.; Hong, S.A.; Ha, H.Y. Performance evaluation of passive DMFC single cells. *J. Power Sources* **2006**, *158*, 1256–1261. [CrossRef]
24. Zainoodin, A.M.; Kamarudin, S.K.; Masdar, M.S.; Daud, W.R.W.; Mohamad, A.B.; Sahari, J. Investigation of MEA degradation in a passive direct methanol fuel cell under different modes of operation. *Appl. Energy* **2014**, *135*, 364–372. [CrossRef]
25. Lai, Q.-Z.; Yin, G.-P.; Zhang, J.; Wang, Z.-B.; Cai, K.-D.; Liu, P. Influence of cathode oxygen transport on the discharging time of passive DMFC. *J. Power Sources* **2008**, *175*, 458–463. [CrossRef]
26. Tang, Y.; Yuan, W.; Pan, M.; Tang, B.; Li, Z.; Wan, Z. Effects of structural aspects on the performance of a passive air-breathing direct methanol fuel cell. *J. Power Sources* **2010**, *195*, 5628–5636. [CrossRef]
27. Chen, R.; Zhao, T.S.; Liu, J.G. Effect of cell orientation on the performance of passive direct methanol fuel cells. *J. Power Sources* **2006**, *157*, 351–357. [CrossRef]
28. Hirano, S.; Kim, J.; Srinivasan, S. High performance proton exchange membrane fuel cells with sputter-deposited Pt layer electrodes. *Electrochim. Acta* **1997**, *42*, 1587–1593. [CrossRef]
29. Cavarroc, M.; Ennadjaoui, A.; Mougenot, M.; Brault, P.; Escalier, R.; Tessier, Y.; Durand, J.; Roualdès, S.; Sauvage, T.; Coutanceau, C. Performance of plasma sputtered fuel cell electrodes with ultra-low Pt loadings. *Electrochem. Commun.* **2009**, *11*, 859–861. [CrossRef]
30. Mougenot, M.; Caillard, A.; Brault, P.; Baranton, S.; Coutanceau, C. High Performance plasma sputtered PdPt fuel cell electrodes with ultra low loading. *Int. J. Hydrogen Energy* **2011**, *36*, 8429–8434. [CrossRef]
31. Gruber, D.; Ponath, N.; Müller, J.; Lindstaedt, F. Sputter-deposited ultra-low catalyst loadings for PEM fuel cells. *J. Power Sources* **2005**, *150*, 67–72. [CrossRef]
32. Fofana, D.; Natarajan, S.K.; Hamelin, J.; Benard, P. Low platinum, high limiting current density of the PEMFC (proton exchange membrane fuel cell) based on multilayer cathode catalyst approach. *Energy* **2014**, *64*, 398–403. [CrossRef]
33. O'hayre, R.; Lee, S.J.; Cha, S.W.; Prinz, F.B. A sharp peak in the performance of sputtered platinum fuel cells at ultra-low platinum loading. *J. Power Sources* **2002**, *109*, 483–493. [CrossRef]
34. Asghari, S.; Shahsamandi, M.H.; Ashraf Khorasani, M.R. Design and manufacturing of end plates of a 5 kW PEM fuel cell. *Int. J. Hydrogen Energy* **2010**, *35*, 9291–9297. [CrossRef]
35. Yoon, S.R.; Hwang, G.H.; Cho, W.I.; Oh, I.H.; Hong, S.A.; Ha, H.Y. Modification of polymer electrolyte membranes for DMFCs using Pd films formed by sputtering. *J. Power Sources* **2002**, *106*, 215–223. [CrossRef]
36. Hamada, K.; Ogawa, T.; Okumura, H.; Ishihara, K.N. The effect of substrate roughness on the properties of RF sputtered AZO thin film. *MRS Commun.* **2019**, *9*, 697–701. [CrossRef]
37. Chang, C.H.; Kryder, M.H. Effect of substrate roughness on microstructure, uniaxial anisotropy, and coercivity of Co/Pt multilayer thin films. *J. Appl. Phys.* **1994**, *75*, 6864–6866. [CrossRef]
38. Cheng, H.; Sun, Y.; Hing, P. The influence of deposition conditions on structure and morphology of aluminum nitride films deposited by radio frequency reactive sputtering. *Thin Solid Films* **2003**, *434*, 112–120. [CrossRef]
39. Ponon, N.K.; Appleby, D.J.R.; Arac, E.; King, P.J.; Ganti, S.; Kwa, K.S.K.; O'Neill, A. Effect of deposition conditions and post deposition anneal on reactively sputtered titanium nitride thin films. *Thin Solid Films* **2015**, *578*, 31–37. [CrossRef]
40. Liu, J.G.; Zhao, T.S.; Liang, Z.X.; Chen, R. Effect of membrane thickness on the performance and efficiency of passive direct methanol fuel cells. *J. Power Sources* **2006**, *153*, 61–67. [CrossRef]

41. Jeong, W.; Chang, I.; Ryu, S.; Zheng, C.; Cha, S.W.; Park, T. Cost-effective and durable Ru-sputtered Pt/C-based membrane–electrode assembly for passive direct methanol fuel cells. *AIP Adv.* **2019**, *9*, 095016. [CrossRef]
42. Zhou, Z.; Wang, S.; Zhou, W.; Jiang, L.; Wang, G.; Sun, G.; Zhou, B.; Xin, Q. Preparation of highly active Pt/C cathode electrocatalysts for DMFCs by an improved aqueous impregnation method. *Phys. Chem. Chem. Phys.* **2003**, *5*, 5485–5488. [CrossRef]
43. Nakagawa, N.; Xiu, Y. Performance of a direct methanol fuel cell operated at atmospheric pressure. *J. Power Sources* **2003**, *118*, 248–255. [CrossRef]
44. Kang, Y.S.; Jung, N.; Choi, K.H.; Lee, M.J.; Ahn, M.; Cho, Y.H.; Sung, Y.E. Anode electrode with carbon buffer layer for improving methanol oxidation reaction in direct methanol fuel cell. *Appl. Surf. Sci.* **2014**, *290*, 246–251. [CrossRef]
45. Okada, M.; Konta, Y.; Nakagawa, N. Carbon nano-fiber interlayer that provides high catalyst utilization in direct methanol fuel cell. *J. Power Sources* **2008**, *185*, 711–716. [CrossRef]

materials

MDPI

Article

Maize Straw as a Valuable Energetic Material for Biogas Plant Feeding

Jakub Mazurkiewicz [1], Andrzej Marczuk [2], Patrycja Pochwatka [3,*] and Sebastian Kujawa [1]

1 Institute of Biosystems Engineering, Poznan University of Life Sciences, Wojska Polskiego 50, Poznań 60-627, Poland; jakub.mazurkiewicz@up.poznan.pl (J.M.); sebastian.kujawa@up.poznan.pl (S.K.)
2 Department of Agricultural, Forestry and Transport Machines, University of Life Sciences in Lublin, Głęboka 28, Lublin 20-612, Poland; andrzej.marczuk@up.lublin.pl
3 Department of Environmental Engineering and Geodesy, University of Life Sciences in Lublin, Leszczyńskiego 7, Lublin 20-069, Poland
* Correspondence: patrycja.pochwatka@up.lublin.pl

Received: 12 September 2019; Accepted: 20 November 2019; Published: 22 November 2019

Abstract: Maize has great potential, especially as a substrate for biofuels production. The aim of this paper is to analyze the possibility of usage in methane fermentation maize straw harvested in different weather conditions, which had an influence on different physical parameters, mainly the dry mass content. The research has shown that maize straw harvested in Central-Eastern Europe can have a broad spectrum of dry mass content, which is related to diverse weather conditions during autumn. However, independently from moisture content, maize straw can be a good (for more wet material) or very good (for more dried straw) substrate for the biogas plant. With the methane productivity reaching 201–207 m^3/Mg of fresh mass, this material is a significantly better substrate than that typically used in Europe maize silage (approximately 105 m^3/Mg FM). It was noted that the retention time for maize straw (36–42 days) is longer than in the case of maize silage (less than 30 days). However, this difference is quite small and can be accepted by the biogas plant operators.

Keywords: maize straw; corn stover; methane production; biogas; substrate

1. Introduction

Maize straw (also known as corn stover) is one of the most common materials produced in agronomy. However, especially in Europe, the utilization of this material has been rather weak. For most farmers, its usage was limited to cutting during harvesting and then ploughing [1]. Removing crop residues from the field for commercial biofuel must balance preventing soil erosion, maintaining soil organic matter, and maintaining or increasing productivity [2,3].

In the last decade, global maize grain production has increased by around 40%, and now amounts to almost 1100 million tonnes [4]. In 2017, EU maize grain production was over 70 million tonnes [5], and for comparison in Beijing in 2012 alone, 1.20 million tonnes maize straw was produced [6], and all over China more than 2.7×10^8 tonnes has been produced annually [7]. Such a high production of maize grain causes crop residues, such as leaves, stalks, husks, and cobs, which can constitute up to 50% of the dry matter yield of whole maize plants [8,9]. The fractions mentioned above have different chemical compositions, structure, and fiber properties [10,11], harvesting times, and even topography or soil types [12,13]. For example, sufficient storage as a result of ensiling operations enables 1.1%–2.2% reduced loss of organic matter compared to in open-air storage (63.1%) [7].

For 1 kg of harvested maize grains (expressed in dry matter), the total mass of maize parts (leaves, cobs, husks, and stalks) is approximatively 1.01 kg of dry matter. Regardless of the characteristics mentioned above, maize straw is a valuable energy resource [9,13,14]. Analysis conducted by scientists from the Poznan University of Life Sciences has shown that annual maize straw production in Poland may reach 4 million tonnes in the near future [15].

1.1. Direct Combustion of Maize Straw

One way of obtaining energy from maize straw is through direct combustion. The calorific values of the maize straw range from 17.65 to 18.6 MJ/kg of dry matter [9,16]. This is the general value given without specifying the proportion of individual fractions, age, moisture content, or variety. The variability of gross energy of different straw fractions during aging was very different. In extreme cases, the differences were even 50% [16]. It is emphasized that such large fluctuations in values were caused by sample heterogeneity and/or lack of consistency in calorimetric procedures, suggesting the adoption of mean values in the range of 16.7–20.9 MJ/kg [16]. In addition, they indicate that the energy content of different maize fractions remains fairly constant over time and between individual plants, and therefore they suggest that in the case of combustion, there is no significant difference regarding what fraction and at what moment the plants were harvested. When comparing maize straw with other biomass fuels, attention should also be paid to the content of ash between 4% and 6.8% (particularly with a high concentration of silica that is more than 34% and potassium levels that are higher than 30%), large amounts of nitrogen (0.6% N), sulfur (0.09% S), and chlorine (0.36% Cl) [9]. These are significantly higher values compared to premium wood pellets [17,18], which must contain less than 0.3% N, 0.05% S, and 0.03% Cl [10].

Research results from other scientists also show quite large variations due to the maize straw fraction and early or late harvest time. The ash content measured was: for husk—2.1%, for cob—1.1% (both for late straw harvest), for leaves from 2.4% to 3.4%, and for stems from 6.0% to 7.3% for the early and late fraction, respectively [13].

Such an ash content can significantly reduce the efficiency of heat exchange as a result of slagging when the combustion temperature is low, and chlorine compounds can accelerate boiler corrosion [19]. From a practical point of view, maize straw showed good caloric potential, but the fuel produced from it would have to be concentrated, and the current combustion technique would require a lot of work due to the large amounts of ash.

1.2. Bioethanol Production

Another way to use the energy stored in maize straw is, for example, the production of bioethanol [18]. In particular, the bioconversion of maize straw has been a subject of interest, because this residue is available in large quantities and at low cost [20,21].

Concern over maize straw removal argued that some must be left on the field to sustain the humic content of the soil and maintain soil productivity, and there may also be constraints in supply and its distribution [3]. In spite of this, maize straw has both economic and environmental potential for the production of bioethanol as a shift from gasoline, and thereby provides a substitute for fossil fuel [22,23].

Maize straw, as a typical lignocellulosic raw material, contains lignin, cellulose, and hemicellulose, which together form a complex polymer structure that limits reaction media or enzymes to close contact with cellulose; in this way, maize straw is not easily converted into bioethanol [24]. For practical reasons, it is characterized by low mass and energy density, is resistant to degradation, and pre-treatment is necessary. Simultaneous saccharification and co-fermentation were developed for the production of cellulosic ethanol, but the concept was misdefined because saccharification and co-fermentation are by no means simultaneous [23]. Lignin is not reactive, which not only takes up the reactor spaces during the enzymatic hydrolysis of the cellulosic component and subsequent fermentation of ethanol, but also requires additional mixing. This significantly impedes the high solid load of lignocellulosic biomass and ethanol titles, which consequently increases the energy consumption for ethanol distillation and discharge stillage, and is another challenge for the production of cellulosic ethanol [23].

Based on 144 data sets that were sufficiently consistent and detailed to take into account the current state of the art of converting maize straw to bioethanol, researchers found that the best technological configurations produced 19%–22% ethanol (dry weight), while the technological configuration with the lowest efficiency produced only 11% ethanol [24]. Besides bioethanol production, it was observed in the review that all technological configurations produced large flows of solid and liquid residues, accounting for 55%–75% of the carbon in the maize straw. Bioethanol from residual maize straw could contribute to lowering CO_2 loads within the transport sector, if used as an amendment to gasoline. The "best-practice", defined as the top 15% cumulative probability with respect to the Global Warming Potential (GWP), suggests that technologies based on steam explosion and ammonia-based pre-treatment statistically appear to be the most promising. These technologies could contribute, with residue energy recovery, to GWP savings of 850–1050 kg CO_{2eq}/Mg dry maize straw solids and produce 178–216 kg of bioethanol [25]. Other analyzes, which are extensive but fairly general, indicate that the greenhouse gas (GHG) savings from cellulosic ethanol in the decentralized system compared to gasoline are 3.35–4.84 Tg CO_2 per year [26].

An interesting development of maize straw processing used for energy production is the method in which the water produced during the pre-treatment stage of bioethanol production was recovered and reused as a co-substrate for the anaerobic digestion. Such water contains high concentrations of lignocellulolytic enzymes, which once added to the fermentation chamber, improved the kinetics of methane production. Compared to digestion in the absence of these additional enzymes, the final methane yield increased by 42%, with a 67% reduction in lag-phase time and a nearly 30% increase in the methane production rate [27].

1.3. Economic and Energy Production Potential

Generally, the cost of agricultural residues for European conditions can be estimated at USD 1–8 GJ. There are large differences in biomass production potential and costs between European regions, with 280 regions (NUTS2). Regions that are distinguished by high potential and low costs include a large part of Poland, the Baltic States, Romania, Bulgaria, and Ukraine. In Western Europe, France, Spain, and Italy are moderately attractive, given the high potential low cost criterion [28].

Research has shown that burning a maize in a furnace can replace a significant amount of fossil fuel traditionally used to dry maize grain in many locations. According to analysis, the use of maize straw as the main source of energy for drying grain would be profitable both for a small scale 8.9 Mg/h dryer with an average of 0.7 MW of heating power, and on a large scale 73 Mg/h dryer, with an average of 6.3 MW [9]. The same authors stated that based on 2004 prices, small-scale drying would have a 14–year payback period with harvest and transport prices of USD 25 Mg/DM (dry matter), and even 8 years if it was dried at higher prices such as USD 45 Mg/DM (due to higher transport costs to the destination) due to a clear impact of the scale effect [9].

Using maize would provide an alternative to maize straw for biochar production. This form of solid biofuel allows for the reduction of transport costs and investments in fuel storage. In addition, the pyrolysis of biomass pellets has obvious advantages, such as lower equipment costs, a simple equipment system, and high production efficiency [29]. Biochar pellets can reduce the costs and problems associated with the transport and storage of biomass due to its stable quality, high resistance to weather conditions, and suitability for long-term storage. For these reasons, some scientists argue that biomass pellets should be used to a much greater extent in the energy sector. At the same time, they add that the popularization and use of pellets is extremely expensive (the cost of pellets is about USD 75/Mg, while the raw material is around USD 3.2/Mg) [29,30].

Based on information from the Slovak Republic, it can be concluded that maize straw in the form of briquettes and pellets has a high energy potential, which is comparable with currently used materials for the production of briquettes and pellets. The authors highly recommended using all current waste materials, as well as all types of waste materials (especially biological) to preserve the main key factors of proper waste management [31].

Newer analyses of the Polish market indicate instead that the transformation of corn straw by methane fermentation to obtain electricity is more profitable than the production of solid fuel from it.

On the one hand, the total amount of energy contained in biogas produced from 1 Mg of maize straw is 9.74 GJ, and this value is slightly lower than the net energy that can be used to burn briquettes [32]. It should be emphasized that over 97% of European biogas plants use biogas for the production of electricity and heat in cogeneration units. Thus, the actual energy value from the energy transformation of maize straw by burning the produced biogas is 0.97 MWh of electricity and 4.21 heat. About 15% of energy is lost, mainly in the form of heat. On the other hand, economic analysis of briquette production shows a much lower profit compared to the consumption of corn straw as a substrate for biogas production. This is mainly due to the fact that the price of heat in Poland is several times lower compared to the price of electricity (in Poland: electric energy price = 150 USD/MWh and heat energy price = 12.5 USD/GJ) [32]. As a consequence, even the conversion of corn straw into solid biofuel (briquettes) can generate more energy from 1 Mg of biomass, thereby providing thermal energy which is cheaper than electricity. The uncertain future of straw use in power plants can be seen in Great Britain. There, changes in energy tariffs caused a decrease in interest from the entire industry in the production of solid fuels from maize straw, as a result of which several plans to build straw granulation plants were suspended [33].

Newer studies analyzing the costs of acquiring and processing maize straw into biogas in more detail indicate that they differ depending on several factors. Maize straw harvesting and storage technology has a significant impact on economic results. Research has shown that collecting corn maize with the help of a self-loading wagon equipped with a cutting system and storing it in a field prism is the most profitable option. The cost of this technology is EUR 12.5 per Mg DM. The cost of harvesting corn straw using a chopper and storage in a flexible silo was the highest, at EUR 126.9 per Mg DM. The test results can be used to estimate the profitability of harvesting corn straw for energy production and other industrial purposes [34].

The use of maize straw for energy purposes is economically and energetically justified. However, the relatively high humidity of this material in its fresh state means that the possibilities of its rational use for heating purposes are strictly limited [35–38]. Therefore, maize straw has far fewer potential buyers, and hence a significantly lower price than straw from cereal ears and even from rapeseed production [36]. However, in the case of methane fermentation, it can be one of the cheaper and more easily available agricultural substrates [39–41].

For example, the total capital investment for the annual volume of ethanol in a decentralized biorefinery is from USD 0.71/dm^3 to USD 1.15/dm^3, and the total capital investment for the annual volume of ethanol production in a centralized biorefinery is USD 1.98/dm^3 [26]. The production costs of renewable diesel is USD 1.19 /dm^3 for the corn stover hydrogen production. The net energy ratio (NER) of the process, which is the ratio of the energy content of the output product to fossil fuel inputs, was calculated to be 1.5 [35]. An interesting comparison was established based on a similitude of the efficiency of algae and maize straw use. Algae are able to achieve the highest surface biomass production rates of 120 Mg/ha·a, ten times higher than maize, but at the same time with the highest costs and the lowest rate of return, in addition to more than three times higher CO_2 emissions levels [42].

It can be concluded from many studies to date that the use of lignocellulosic ethanol (as in the case of maize straw) on a large scale would require more sustainable and cost-effective agricultural practices, since ethanol production is highly dependent on the cost of the raw material, as well as the development of advanced ethanol biorefinery technologies [43,44]. Generally, the price of raw material has a big impact on raw material supply chains, although not directly. Namely, the higher raw material price has no major impact on the final ethanol sales price, but higher raw material prices significantly increase the amount of ethanol produced. Farmers significantly increase the supply of cellulose biomass at higher prices of raw materials, then the reduction of the final price is primarily due to transport costs and the effect of scale that larger biorefineries have, thus reducing the ethanol sales price [44,45]. Currently, only single studies can be found that suggests that producing bioethanol at smaller local points is economically profitable [26].

1.4. Biogas Production

There are almost 20,000 biogas plants working in Europe. Except for in Denmark, Switzerland, and Sweden, the most popular substrate used for feeding agricultural biogas plants is maize silage produced from whole plants. This substrate maintains stable biogas production on a good level (approximately 105 m^3 CH_4/Mg of Fresh Mass) and is easy to harvest and store. With a high level of subsidies in many EU countries, this substrate (despite being quite expensive) has guaranteed good rentability for most of the agricultural biogas plants. That was a major reason for the huge development of maize cultivation for biogas plants feeding in places like Germany where over 10% of the agricultural land surface is used for silage production to cover biogas sector needs.

However, during last decade, many European countries have introduced limitations for Renewable Energy Sources subsidies. This also affected the biogas sector. From the other side, maize silage prices experienced significant growth, which reduced biogas plant rentability. That is why many biogas plant operators started to look for alternative substrates which can be easy to get, are inexpensive, and have good methane productivity. Maize straw can be one of these substrates, which has also been confirmed by Chinese experiences [7,46].

The aim of this paper is to analyze the possibility of using maize straw harvested in different weather conditions in a methane fermentation process, and as a consequence having different physical parameters (mainly different dry mass content). It has to be mentioned that weather conditions in Central-Eastern Europe during autumn (the time of maize harvesting for grain production) are very unstable. That is why maize straw parameters are deeply dependent on weather conditions.

2. Materials and Methods

The analyzed material (different maize straws) was collected from 4 farms situated in Western and Eastern Poland. Straw was collected manually after a grain harvest (average 10 kg for each sample), directly from the field, and then transported to the Ecotechnologies Laboratory (at the Poznan University of Life Sciences, PULS) for basic analyses and biogas production tests using methane fermentation. Straw was collected under different weather conditions, which should have a strong influence on its physical characteristics.

Collected materials (different maize straws) were classified within dry matter content. That is why the material names (each maize straw, MS) contain dry matter (TS) content. The analyzed straws names were: MS45, MS55, MS78, and MS89 (numbers mean the initial dry matter content is expressed based on the percentage of dry matter).

2.1. Physical Analysis

Maize samples initially were analyzed for dry matter content (Total Solids—TS, within the Polish Standard PN-75 C-04616/01 [47]). The standard procedure contains drying samples (in 3 repetitions) heated to 105 °C over 24 h. Organic matter (Volatile Solids, VS) content was analyzed within the standard PN-Z-15011-3 [48] by combustion of dry samples (in 3 repetitions) at 525 °C over 3 h. The samples were also analyzed for pH (PN-90 C-04540/01 [49]) and conductivity (PN-EN 27888: 1999 [50]).

It has to be underlined that data about TS and VS are indispensable for starting fermentation tests and then, for calculating the methane and biogas production efficiency of each analyzed straw into the units m^3/Mg FM (Fresh Matter), m^3/Mg TS, and m^3/Mg VS. While results expressed in CH_4 production in m^3 per VS are most commonly used in research papers to compare the efficiency of each substrate, in real biogas plant exploitation the most important parameter is substrate methane efficiency expressed in CH_4 by Mg of FM (Fresh Matter).

2.2. Methane Fermentation Tests

Biogas production analysis was conducted at Ecotechnologies Laboratory, the biggest Polish biogas unit. The laboratory uses the German's standards: DIN 38 414/S8 [51] and VDI 4630 [52].

This laboratory received the Proficiency Test Biogas certificate for a framework of international tests organized by German KTBL in 2017, as it was the first Polish biogas laboratory which provided high-quality research in the area of methane fermentation. The experiment of straw samples anaerobic digestion has been conducted in a special 21-chambers fermenter. The simple version of a 3-chambers section fermenter is shown in Figure 1.

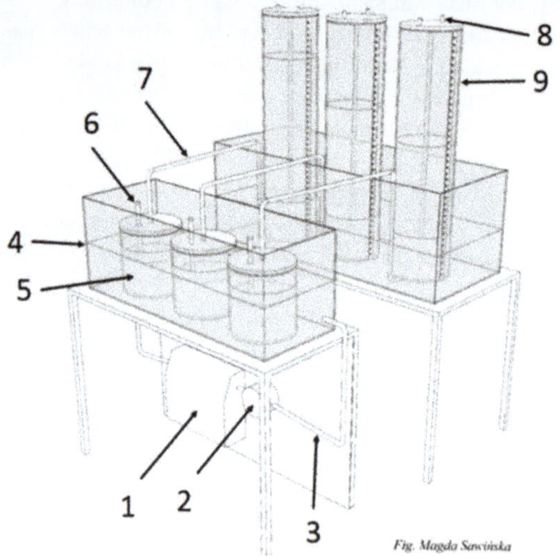

Figure 1. Scheme of a fermenter for biogas production research (3-chamber section): 1—water heater with temperature regulator, 2—water pump, 3—insulated conductors of calefaction liquid, 4—water coat, 5—fermenter with charge capacity 2 dm^3, 6—sampling tubes, 7—biogas transporting tube, 8 - gas sampling valve, 9—biogas volume-scale reservoir [53].

The fermentation procedure of the anaerobic digestion tests proceeded in glass fermenters with volumes of 2 dm^3 each. The analyzed substrates were put inside each reactor with the weight-related to the VS content and then were mixed with inoculum standardized with DIN 38 414/S8 and VDI 4630 standards. Reactors were flushed with nitrogen gas in order to remove air and immediately create anaerobic conditions (oxygen is an inhibitor for methanogenic bacteria). All samples were fermented in 3 replications, with final results representing an average value of those measurements.

The biogas production measurements were taken every 24 h by connecting gas reservoirs (9) (Figure 1) through sampling valve (8) with gas analyzer Geotech GA5000. This analyzer possesses quality certifications like ATEX II 2G Ex ib IIA T1 Gb (Ta = −10 °C to + 50 °C), IECEx, CSA, and the calibration certificate UKAS ISO 17025. The measurement ranges for the Geotech analyzer are: O_2 0–25%, CO_2 0–100%, CH_4 0–100%, NH_3 0–1000 ppm, and H_2S 0–10 000 ppm. GA5000 was usually calibrated once a week using calibration gases (from the Air Product company) in the concentrations: 65% for CH_4, 35% for CO_2 (in one mixture), 500 ppm H_2S, 100 ppm NH_3, and O_2 using synthetic air.

The end of the methane fermentation experiment (within DIN 38 414/S8 standard) was the moment when daily biogas production was less than 1% of the total volume of obtained biogas. The efficiency of examined materials for biogas and methane productivities can be expressed using low, good, or very good levels. In described research, this criterion concerns chemical energy production from maize straw and is expressed in the amount of methane produced from the mass unit (m^3 CH_4/Mg). Since the most common substrate used in European biogas plants is maize silage with the methane productivity 105 m^3/Mg, all straws was analyzed under the criteria "low quality substrate", "good substrate" (like maize silage), and "very good substrate".

3. Results

3.1. Physical Analysis

The analysis of dry matter (TS) and organic dry matter (VS) is shown in Table 1.

Table 1. The initial parameters of the tested material (maize straws) used for fermentation.

Substrate	TS (% FM)	VS (% TS)
MS45	44.77	97.59
MS55	54.68	93.41
MS78	77.80	89.78
MS89	89.46	89.50

The analysis of TS shows a significant difference between tested materials because dry matter content varies from 45% to 89%. This vast difference is related to very unstable weather conditions (different rainfall and temperatures) during the autumn period in Poland and as a consequence strongly influences the viability of maize straw storage technologies. It has to be underlined that for more dry materials like in case of MS78 and MS89, maize straw can be stored in the pressed form (bales). However, a material with higher moisture like MS45 and MS55 should be stored over a longer time as silage because the additional tests of more wet straw storage in bales increase their temperature and start the quasi composting process. This phenomena strongly reduces the energetic potential of maize straw through non-controlled losses of heat from worm bales.

Considerably smaller differences between analyzed materials were found in the case of organic matter content (VS). The highest VS content in the straw with higher moisture content could be related with the "cleaning" effect of the rains that occurred before and during the harvest. The straws harvested during dry weather contained more soil dust which increased mineral matter content up to over 10% in case of MS78 and MS89 (Table 1). In contrast, the most wet straw (MS45) harvested under rainy weather only had 2.41% mineral matter and the highest organic matter content (97.59%).

3.2. Fermentation Results

The results of biogas production are presented on Figure 2. It is important to underline that results of biogas production (expressed in dm^3) cannot be directly compared at this stage between tested materials because of the different initial fresh mass used for tests. The German standard DIN 38 414/S8 needs to use the amount of sample calculated in organic dry matter (VS), so the initial fresh matter amount is usually different. The final biogas and methane productivity results (after calculations required by the DIN standard) are presented in Table 2.

Table 2. The biogas and methane production from tested materials.

Substrate	CH_4 Content (%)	CH_4 m^3/ Mg FM	Biogas m^3/ Mg FM	CH_4 m^3/ Mg TS	Biogas m^3/ Mg TS	CH_4 m^3/ Mg VS	Biogas m^3/ Mg VS
MS45	49.51	103.39	206.47	230.93	461.17	236.64	472.56
MS55	48.97	98.95	199.94	180.96	366.09	193.73	391.45
MS78	50.17	201.12	400.90	258.52	515.29	287.94	573.95
MS89	50.26	206.78	411.43	231.14	459.90	258.26	513.86

The results of the biogas efficiency test show that the most intensive gas production occurred during the first 10 days of the process. The fermentation time was almost the same for 3 analyzed materials (36–37 days), while only the driest straw (MS89) had the process occur for 42 days. This was probably related to having the highest concentration of hardly fermentable fibers and can be a subject of future study.

However, the methane production results are shown in Figure 3.

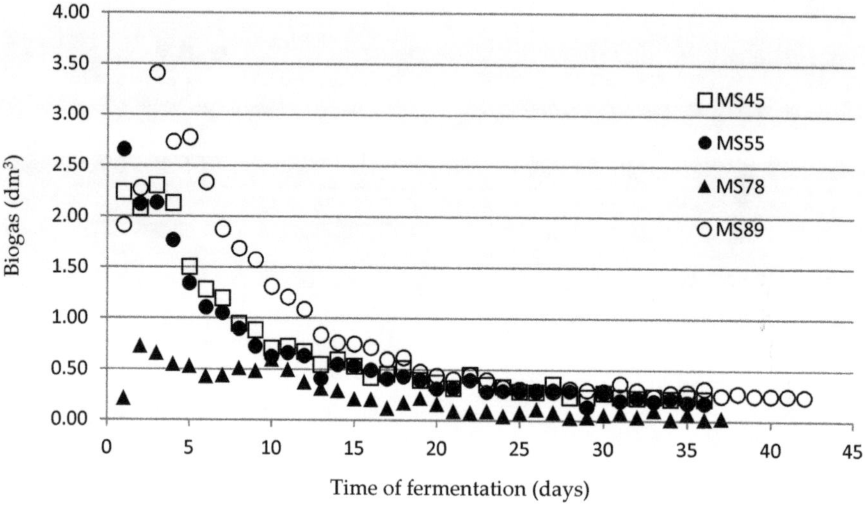

Figure 2. Production of biogas during materials fermentation process (daily measurements) for all materials.

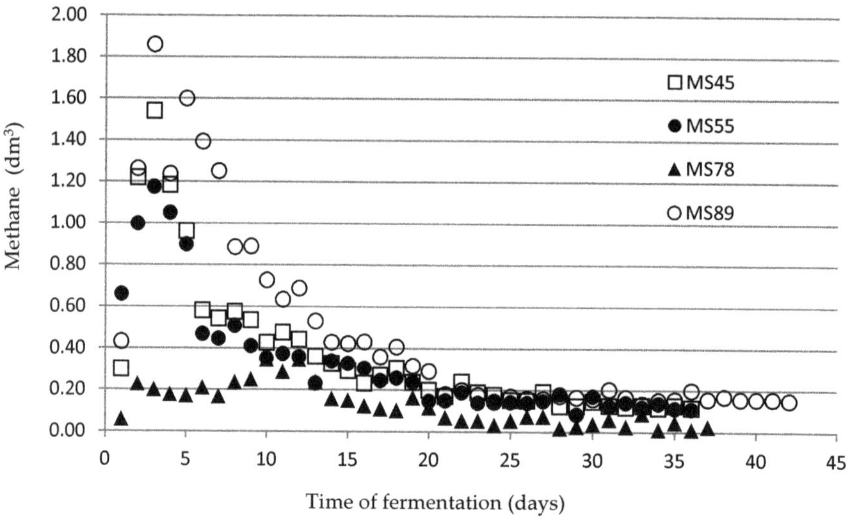

Figure 3. Methane production during fermentation process (daily measurements) for all materials.

It should be underlined that methane (contained in biogas) is the crucial parameter of estimated substrates. The biogas contains mainly methane (the fuel for the co-generation unit) and carbon dioxide (the ballast). Similarly to for biogas production, methane generation also occurred mainly during the first 10–13 days. After the 20th day, the methane production stayed low for all tested materials.

Using the procedures described in the standard DIN 38 414/S8, the calculations were made in order to present the biogas and methane production results from analyzed straws, expressed in the unit which is the most often used by specialists from the biogas sector (m^3 of methane from Mg of the substrate). Those calculations also take into account the production from inoculum used for initializing the fermentation process. The results of biogas and methane productivity from tested straws are shown in Table 2.

The data content in Table 2 shows that methane concentration was very similar (approximately 49%–50%) for all tested materials. The CH_4 production from analyzed material was very different (from 99 to 207 m^3/Mg) in the calculation based on the fresh matter, but this is logical because a higher concentration of dry matter correlates with higher methane production. Typical maize silage from whole plants can reach the methane's productivity level at approximatively 105 m^3/Mg. The results obtained for wet straws (MS45 and MS55) are a little bit lower compared to maize silage, but in case of more dry straws, the productivity is almost 2 times higher.

However, when looking at CH_4 production expressed in m^3/Mg of VS, the differences in productivity level are much lower (194–288 m^3/Mg VS). This means that independently of maize straw moisture, this material is still attractive as a substrate for a biogas plant.

From a biogas plant manager's point of view, the time of substrate total fermentation is very important because with shorter biogas production process more substrate can be treated in fermenters with the same volume. Figure 4 shows the time needed for complete fermentation as well as for 80% and 90% methane production levels.

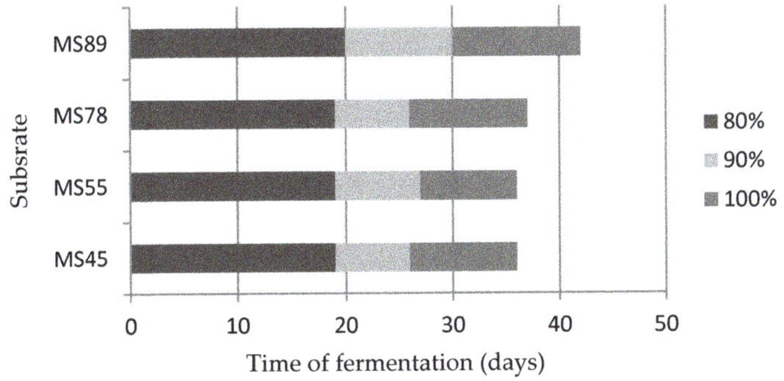

Figure 4. Time required 80%, 90%, and 100% total methane production from analyzed maize straws.

The time required for complete fermentation of analyzed straws is between 36 and 42 days. However, 90% of methane production occurred before 30 days of the process. This means that a shorter process in the main fermenters (up to 30 days) can generate most of the energetic gas. The rest can be obtained in the digested tank, if it is covered by a hermetic cover and is equipped with a heating system. It has to be underlined that the time needed for complete methanization of all analyzed straws was longer than for a typical maize silage fermentation, which is usually less than 30 days [53]. However, maize straw retention time (36–42 days) is still acceptable for biogas plant operators and is almost 3 times shorter than for a typical straw fermentation (105–120 days) in NaWaRo technology, which is the most frequently used technology type in Europe.

4. Discussion

Maize straw is a material produced in huge quantities on a worldwide scale. However, compared to cereals straw, this material is rarely used in Europe and usually stays on the field after harvest as an organic fertilizer [54]. In this study, we have shown that maize straw can be a valuable substrate for a biogas plant, especially if the material has a higher dry matter content. In this case, the methane productivity (201–207 m^3/Mg FM in the case of MS78 and MS89) is clearly higher compared to maize silage from whole plants (approximately 105 m^3/Mg FM), which is the most popular substrate in European biogas plants [46,55–57].

One of the problems related to maize straw utilization is farmer fear of proper methods for its harvest and storage. However, within the PULS research project conducted in 2010–2013 by the scientists from Institute of Biosystems Engineering, 6 harvest technologies and 3 maize straw storage technologies were developed. This and the research that followed have shown the large possibilities of maize straw harvesting and utilization for biogas plants feeding in the form of silage, but also directly as dry material stored in bales [34,57,58]. The farmers should therefore be able to know the moisture content of harvested maize straw. However, this parameter should be the key element for making a decision about the maize straw storage technology they should use: either using compressed bales or a silage process under the chopped straw form. Taking into account the possible technologies of maize straw management, we think that weather conditions can be important during harvesting time would not prevent maize straw utilization as a valuable substrate for biogas plant. In this case, the farmers should measure the dry matter content/moisture of maize straw before harvest and then choose the best technology for its storage (in bales or in silos).

The significant advantage of maize straw (instead of maize silage) usage for biogas is the exclusion of the conflict between food and biofuels production. Maize straw is the additional biomass on the field from grain production. Its utilization for biogas production helps to manage the chemical energy contained in straw organic matter, which under typical scenarios is used by microorganisms proceeding its decomposition in soil. It is also important the macro and microelements contained in maize straw (except carbon) are not lost in the case of using the straw in a biogas plant because, after fermentation, those minerals came back to the soil with digestate.

There are some ideas to use the maize straw for bioethanol production, solid biofuels, or as a structural substrate for composting in windrows [32,38,58]. However after taking energetic, economic, and agricultural into account aspects, the usage of maize straw as a substrate for biogas plants seems to be the most promising option for the near future.

5. Conclusions

The research described in this paper let us reach the following conclusions:

1. Maize straw harvested in Central-Eastern Europe can have a broad spectrum of dry mass content, which is related to diverse weather conditions during autumn.
2. Independently from moisture content, maize straw can be good (for more wet material) or very good (for more dried straw) substrate for the biogas plant. With the methane productivity reaching 201–207 m^3/Mg of fresh mass, this material is significantly better substrate than that typically used in Europe maize silage (approximately 105 m^3/Mg FM).
3. The retention time for maize straw (36–42 days) is longer than for maize silage (less than 30 days). However, this difference is quite small and can be accommodated by the biogas plant operators.

Author Contributions: Conceptualization, J.M. and P.P.; methodology, J.M. and S.K.; software, A.M.; formal analysis, S.K.; investigation, P.P., and S.K.; resources, J.M. and A.M.; data curation, P.P.; writing—original draft preparation, J.M.; writing—review and editing, J.M., P.P. and S.K.; visualization, P.P.; supervision, J.M.; project administration, A.M. and P.P.; funding acquisition, A.M.

Funding: This research received no external funding.

Conflicts of Interest: The authors declare no conflict of interest.

References

1. Dach, J.; Professor at Poznan University of Life Sciences, Poznań, Poland. Personal communication, June 2019.
2. Johnson, J.; Reicosky, D.; Sharratt, B.; Lindstrom, M.; Voorhees, W. Corn Stover as a Biofuel. Available online: https://www.researchgate.net/publication/252816491_Corn_stover_as_a_biofuel (accessed on 26 August 2019).
3. Richard, T.L. Challenges in Scaling Up Biofuels Infrastructure. *Science* **2010**, *329*, 793–796. [CrossRef] [PubMed]
4. Grain Production Worldwide by Type, 2018/19|Statista. Available online: https://www.statista.com/statistics/263977/world-grain-production-by-type/ (accessed on 26 August 2019).
5. Agricultural Production—Crops—Statistics Explained. Available online: https://ec.europa.eu/eurostat/statistics-explained/index.php/Agricultural_production_-_crops (accessed on 26 August 2019).

6. Wandera, S.M.; Qiao, W.; Algapani, D.E.; Bi, S.; Yin, D.; Qi, X.; Liu, Y.; Dach, J.; Dong, R. Searching for possibilities to improve the performance of full scale agricultural biogas plants. *Renew. Energy* **2018**, *116*, 720–727. [CrossRef]
7. Sun, H.; Cui, X.; Stinner, W.; Mustafa Shah, G.; Cheng, H.; Shan, S.; Guo, J.; Dong, R. Synergetic effect of combined ensiling of freshly harvested and excessively wilted maize stover for efficient biogas production. *Bioresour. Technol.* **2019**, *285*, 121338. [CrossRef] [PubMed]
8. Czajkowski, Ł.; Wojcieszak, D.; Olek, W.; Przybył, J. Thermal properties of fractions of corn stover. *Constr. Build. Mater.* **2019**, *210*, 709–712. [CrossRef]
9. Morissette, R.; Savoie, P.; Villeneuve, J. Combustion of Corn Stover Bales in a Small 146-kW Boiler. *Energies* **2011**, *4*, 1102–1111. [CrossRef]
10. Barten, T.J. Evaluation and Prediction of Corn Stover Biomass and Composition from Commercially Available Corn Hybrids. Ph.D. Thesis, Iowa State University, Ames, IA, USA, 2013.
11. Wang, J.S.; Steinberger, Y.; Wang, X.Y.; Hu, L.; Chen, X.; Xie, G.H. Variations of Chemical Composition in Corn Stover Used for Biorefining. Available online: https://www.ingentaconnect.com/content/asp/jbmb/2014/00000008/00000006/art00010 (accessed on 26 August 2019).
12. Berchem, T.; Roiseux, O.; Vanderghem, C.; Boisdenghien, A.; Foucart, G.; Richel, A. Corn stover as feedstock for the production of ethanol: Chemical composition of different anatomical fractions and varieties. *Biofuels Bioprod. Biorefin.* **2017**, *11*, 430–440. [CrossRef]
13. Garlock, R.J.; Chundawat, S.P.; Balan, V.; Dale, B.E. Optimizing harvest of corn stover fractions based on overall sugar yields following ammonia fiber expansion pretreatment and enzymatic hydrolysis. *Biotechnol. Biofuels* **2009**, *2*, 29. [CrossRef]
14. Dupont, C.; Chiriac, R.; Gauthier, G.; Toche, F. Heat capacity measurements of various biomass types and pyrolysis residues. *Fuel* **2014**, *115*, 644–651. [CrossRef]
15. Wojcieszak, D.; Przybył, J.; Dach, J.; Zaborowicz, M.; Staszak, Ż. Economic Assessment of the Technology Harvesting Maize Straw for Biogas Production. *BIO Web Conf.* **2018**, *10*, 01017. [CrossRef]
16. Pordesimo, L.O.; Hames, B.R.; Sokhansanj, S.; Edens, W.C. Variation in corn stover composition and energy content with crop maturity. *Biomass Bioenergy* **2005**, *28*, 366–374. [CrossRef]
17. Kowalczyk-Juśko, A.; Kowalczuk, J.; Szmigielski, M.; Marczuk, A.; Jóźwiakowski, K.; Zarajczyk, K.; Masłowski, A.; Ślaska-Grzywna, B.; Sagan, A.; Zarajczyk, J. Quality of biomass pellets used as fuel or raw material for syngas production. *Przemysl Chem.* **2015**, *94*, 1835–1837. [CrossRef]
18. Wiśniewski, D.; Piechocki, J.; Białowiec, A.; Pulka, J.; Siudak, M.; Jakubowski, B.; Myślak, B. Operational studies of prototype biomass gasification reactor. *Rocz. Ochr. Srodowiska* **2015**, *17*, 1094–1112.
19. Kowalczyk-Juśko, A.; Kościk, B.; Jóźwiakowski, K.; Marczuk, A.; Zarajczyk, J.; Kowalczuk, J.; Szmigielski, M.; Sagan, A. Effects of biochemical and thermochemical conversion of sorghum biomass to usable energy. *Przemysl Chem.* **2015**, *94*, 1000–1002. [CrossRef]
20. Wilhelm, W.W.; Johnson, J.M.F.; Karlen, D.L.; Lightle, D.T. Corn Stover to Sustain Soil Organic Carbon Further Constrains Biomass Supply. *Agron. J.* **2007**, *99*, 1665–1667. [CrossRef]
21. Das, S.; Peterson, B.; Chin, S.-M. Analysis of Fuel Ethanol Transportation Activity and Potential Distribution Constraints. *Transp. Res. Rec.* **2010**, *2168*, 136–145. [CrossRef]
22. Morales, M.; Quintero, J.; Conejeros, R.; Aroca, G. Life cycle assessment of lignocellulosic bioethanol: Environmental impacts and energy balance. *Renew. Sustain. Energy Rev.* **2015**, *42*, 1349–1361. [CrossRef]
23. Zabed, H.; Sahu, J.N.; Boyce, A.N.; Faruq, G. Fuel ethanol production from lignocellulosic biomass: An overview on feedstocks and technological approaches. *Renew. Sustain. Energy Rev.* **2016**, *66*, 751–774. [CrossRef]
24. Zhao, Y.; Damgaard, A.; Christensen, T.H. Bioethanol from corn stover—A review and technical assessment of alternative biotechnologies. *Prog. Energy Combust. Sci.* **2018**, *67*, 275–291. [CrossRef]
25. Zhao, Y.; Damgaard, A.; Xu, Y.; Liu, S.; Christensen, T.H. Bioethanol from corn stover—Global warming footprint of alternative biotechnologies. *Appl. Energy* **2019**, *247*, 237–253. [CrossRef]
26. Kim, S.; Zhang, X.; Dale, B.E.; Reddy, A.D.; Jones, C.D.; Izaurralde, R.C. EISA (Energy Independence and Security Act) compliant ethanol fuel from corn stover in a depot-based decentralized system. *Biofuels Bioprod. Biorefin.* **2018**, *12*, 873–881. [CrossRef]

27. Jazmín Edith, M.-H.; Octavio, L.; Edna Madai, M.-H.; Esperanza, H.; Nicolás Óscar, S.-C. Coupling energy-production processes: The use of residues from bioethanol production to improve the anaerobic digestion of corn stover. *Biomass Bioenergy* **2019**, *128*, 105322. [CrossRef]
28. De Wit, M.; Faaij, A. European biomass resource potential and costs. *Biomass Bioenergy* **2010**, *34*, 188–202. [CrossRef]
29. Xing, X.; Fan, F.; Jiang, W. Characteristics of biochar pellets from corn straw under different pyrolysis temperatures. *R. Soc. Open Sci.* **2018**, *5*, 172346. [CrossRef] [PubMed]
30. Yin, W.; Guo, Y.; Hu, F.; Fan, Z.; Feng, F.; Zhao, C.; Yu, A.; Chai, Q. Wheat-Maize Intercropping with Reduced Tillage and Straw Retention: A Step Towards Enhancing Economic and Environmental Benefits in Arid Areas. *Front. Plant Sci.* **2018**, *9*. [CrossRef] [PubMed]
31. Urbanovičová, O.; Angelovič, M.; Križan, M.; Krištof, K.; Jobbágy, J. Energy potential of densified biomass from maize straw in form of pellets and briquettes. *Agron. Res.* **2018**, *16*, 474–482. [CrossRef]
32. Zbytek, Z.; Dach, J.; Pawłowski, T.; Smurzyńska, A.; Czekała, W.; Janczak, D. Energy and economic potential of maize straw used for biofuels production. In *MATEC Web of Conferences*; EDP Sciences: Les Ulis, France, 2016; Volume 60. [CrossRef]
33. Straw Burning Plants Contributing to "Severe Straw Shortages". Available online: http://www.npa-uk.org.uk/Straw_burning_plants_contributing_to_severe_straw_shortages.html (accessed on 31 October 2019).
34. Wojcieszak, D.; Przybył, J.; Mazurkiewicz, J.; Janczak, D.; Zaborowicz, M. Increasing the energy value of corn stover used in biogas plant without pre-processing. In Proceedings of the 18th International Multidisciplinary Scientific GeoConference SGEM2018, Sofia, Bulgaria, 3–6 December 2018; Volume 18, pp. 495–502. [CrossRef]
35. Haseli, Y. Simplified model of torrefaction-grinding process integrated with a power plant. *Fuel Process. Technol.* **2019**, *188*, 118–128. [CrossRef]
36. Womac, A.R.; Igathinathane, C.; Sokhansanj, S.; Pordesimo, L.O. Biomass moisture relations of an agricultural field residue: Corn stover. *Trans. Am. Soc. Agric. Eng.* **2005**, *48*, 2073–2083. [CrossRef]
37. Kaliyan, N.; Morey, R.V. Densification characteristics of corn stover and switchgrass. *Trans. ASABE* **2009**, *52*, 907–920. [CrossRef]
38. Kujawa, S.; Nowakowski, K.; Tomczak, R.J.; Dach, J.; Boniecki, P.; Weres, J.; Mueller, W.; Raba, B.; Piechota, T.; Rodríguez Carmona, P.C. Neural image analysis for maturity classification of sewage sludge composted with maize straw. *Comput. Electron. Agric.* **2014**, *109*, 302–310. [CrossRef]
39. Czekała, W.; Dach, J.; Janczak, D.; Smurzyńska, A.; Kwiatkowska, A.; Kozłowski, K. Influence of maize straw content with sewage sludge on composting process. *J. Water Land Dev.* **2016**, *30*, 43–49. [CrossRef]
40. Dobek, T.; Šařec, P.; Kołosowski, P.; Šařec, O. Effect of chopped maize straw on the quantity and quality of biogas produced. *Res. Agric. Eng.* **2017**, *63*, S8–S12. [CrossRef]
41. Smurzyska, A.; Dach, J.; Kozowski, K.; Mazurkiewicz, J.; Woniak, E.; Boniecki, P.; Kupryaniuk, K.; Janczak, D.; Brzoski, M. Relevant biogas substrate—Maize silage vs slaughterhouse waste. In Proceedings of the International Conference on Information and Communication Technologies in Agriculture, Food and Environment, Chania, Crete Island, Greece, 21–24 September 2017; Volume 2030, pp. 196–204.
42. Christiansen, K.; Raman, D.R.; Hu, G.; Anex, R. First-order estimates of the costs, input-output energy analysis, and energy returns on investment of conventional and emerging biofuels feedstocks. *Biofuel Res. J.* **2018**, *5*, 894–899. [CrossRef]
43. González-García, S.; Moreira, M.T.; Feijoo, G. Comparative environmental performance of lignocellulosic ethanol from different feedstocks. *Renew. Sustain. Energy Rev.* **2010**, *14*, 2077–2085. [CrossRef]
44. Kim, S.; Dale, B.E. All biomass is local: The cost, volume produced, and global warming impact of cellulosic biofuels depend strongly on logistics and local conditions. *Biofuels Bioprod. Biorefin.* **2015**, *9*, 422–434. [CrossRef]
45. Baral, N.R.; Quiroz-Arita, C.; Bradley, T.H. Uncertainties in corn stover feedstock supply logistics cost and life-cycle greenhouse gas emissions for butanol production. *Appl. Energy* **2017**, *208*, 1343–1356. [CrossRef]
46. Yu, Q.; Liu, R.; Li, K.; Ma, R. A review of crop straw pretreatment methods for biogas production by anaerobic digestion in China. *Renew. Sustain. Energy Rev.* **2019**, *107*, 51–58. [CrossRef]
47. Polish Standard PN-75 C-04616/01. *Determination of Dry Matter of Sludge and Organic Substances. Water and Sewage. Special Sludge Tests. Determination of Water Content, Dry Matter, Organic Substances and Mineral Substances in Sewage Sludge*; Polish Committee for Standardization: Warsaw, Poland, 1975.

48. Polish Standard PN-Z-15011-3. *Municipal Waste Compost—Determination of pH, Organic Matter Content, Organic Carbon, Nitrogen, Phosphorus and Potassium*; Polish Committee for Standardization: Warsaw, Poland, 2001.
49. Polish Standard PN-90 C-04540/01. *Water and Waste Water—Tests for pH, Acidity and Alkalinity. Electrometric pH Determination of Water and Waste Water with Specific Conductivity 10 μS/cm and Higher*; Polish Committee for Standardization: Warsaw, Poland, 1990.
50. Polish Standard PN-EN 27888: 1999. *Water Quality—Determination of Specific Electrical Conductivity*; Polish Committee for Standardization: Warsaw, Poland, 1999.
51. DIN 38 414/S8. *German Standard Methods for the Examination of Water, Waste Water and Sludge; Sludge and Sediments (Group S); Determination of the Amenability to Anaerobic Digestion (S 8)*; DIN Deutches Institut für Normung e. V.: Berlin, Germany, 2012.
52. VDI 4630. *Fermentation of Organic Materials. Characterization of the Substrate, Sampling, Collection of Material Data, Fermentation Tests*; Verein Deutscher Ingenieure e.V.: Düsseldorf, Germany, 2016.
53. Cieślik, M.; Dach, J.; Lewicki, A.; Smurzyńska, A.; Janczak, D.; Pawlicka-Kaczorowska, J.; Boniecki, P.; Cyplik, P.; Czekała, W.; Jóźwiakowski, K. Methane fermentation of the maize straw silage under meso- and thermophilic conditions. *Energy* **2016**, *115*, 1495–1502. [CrossRef]
54. Dach, J.; Jóźwiakowski, K.; Kowalczyk-Juśko, A.; Kozłowski, K.; Neugebauer, M. Biogas plant exploitation under the low-subsidies market conditions: Maize silage versus biowaste scenarios. In Proceedings of the 16th International Multidisciplinary Scientific GeoConference SGEM 2016, Sofia, Bulgaria, 2–5 November 2016.
55. Janczak, D.; Kozłowski, K.; Zbytek, Z.; Cieślik, M.; Bugała, A.; Czekała, W. Energetic efficiency of the vegetable waste used as substrate for biogas production. In *MATEC Web of Conferences*; EDP Sciences: Les Ulis, France, 2016; Volume 64. [CrossRef]
56. Kozłowski, K.; Dach, J.; Czekała, W.; Lewicki, A. Influence of Maize Silage Storage Conditions on Biogas Efficiency. Available online: https://www.researchgate.net/publication/308994841_Influence_of_maize_silage_storage_conditions_on_biogas_efficiency (accessed on 11 September 2019).
57. Kozłowski, K.; Mazurkiewicz, J.; Chełkowski, D.; Jeżowska, A.; Cieślik, M.; Brzoski, M.; Smurzyńska, A.; Dongmin, Y.; Wei, Q. The effect of mixing during laboratory fermentation of maize straw with thermophilic technology. *J. Ecol. Eng.* **2018**, *19*, 93–98. [CrossRef]
58. Przybył, J.; Wojcieszak, D.; Kowalik, I.; Dach, J. Influence of the Harvesting and Ensilage Technology on the Quality of Maize Straw Silage. *BIO Web Conf.* **2018**, *10*, 02027. [CrossRef]

materials

MDPI

Article

Characterization of Corrosion Behavior of CLF-1 in Liquid Lithium Using Calibration-Free Laser-Induced Breakdown Spectroscopy in Depth Profile Analysis

Zhi Cao [1,2], Yongtao An [2,*], Xianglin Wang [2], Chang'an Chen [2] and Ying Li [3]

1 Key Laboratory of Radiation Physics and Technology, Ministry of Education, Institute of Nuclear Science and Technology, Sichuan University, Chengdu 610065, China; georgecao7@gmail.com
2 Institute of Materials, China Academy of Engineering Physics, Jiangyou 621908, China; terno@126.com (X.W.); chenchangan@caep.cn (C.C.)
3 Key Laboratory of Advanced Technology of Materials, Ministry of Education, Superconductivity and New Energy Research and Development Center, Southwest Jiaotong University, Chengdu 610031, China; jierze@163.com
* Correspondence: anyt03@163.com

Received: 11 December 2019; Accepted: 2 January 2020; Published: 6 January 2020

Abstract: It is important to get fast and quantitative compositional depth profiles for the boundary layer of the corroded specimen in order to understand the corrosion process and mechanism due to liquid lithium induced corrosion problems to structural material of fusion reactors. In this work, calibration-free laser-induced breakdown spectroscopy (CF-LIBS) is introduced to investigate the compatibility of CLF-1(China low-activation Ferritic steel) exposed in liquid lithium at 500 °C for 500 h. The results show that CF-LIBS constitutes an effective technique to observe the corrosion layer of specimens which are non-uniform and the elements of matrix show gradient distribution from the boundary to the inner layer. The concentration was 82–95 wt.% Fe, 5–12 wt.% Cr, 0.45–0.85 wt.% Mn, 1.6–1.1 wt.% W, 0.11–0.16 wt.% V, and <0.2 wt.% Li along the longitudinal corrosion depth for the corrode CLF-1. The results reveal the quantitative elemental variation trend of CLF-1 in the lithium corrosion process and indicate that the CF-LIBS approach can be applied to the analysis of composition in multi-element materials.

Keywords: lithium corrosion; calibration-free laser-induced breakdown spectroscopy; quantitative analysis; depth profile analysis

1. Introduction

Liquid lithium, as the self-coolant and the tritium breeder applied in Tokamak reactor, has the advantage of high atom density, high heat conductivity. Moreover, its liquid phase could largely avoid the problem caused by thermal expansion and cold shrinkage, as well as radiation damage [1,2]. Thus, the demonstration of stability of structural material against corrosion caused by lithium is crucial [3,4]. There are several major candidate materials for the blanket structure, such as reduced activation ferritic-martensitic (RAFM) [3–5], China low-activation martensitic (CLAM) [6], and vanadium-allay. A kind of Chinese RAFM named CLF-1 fabricated by the Southwestern Institute of Physics (SWIP) has good mechanical properties and low sensitivity to radiation-induced swelling, and is considered the structural material candidate for the China Fusion Engineering Test Reactor (CFETR) [7].

The core corrosive problem of structural material against liquid lithium mainly involves the dissolution and penetration of elements [3,8]. In order to investigate the dissolution and penetration of the lithium corrosion process, several technical methods have been considered, such as secondary ion mass spectrometry (SIMS), X-ray photoelectron spectroscopy (XPS), the glow discharge optical emission spectroscopy & mass spectrometry (GD-OES/MS), and inductively coupled plasma mass spectrometry and atomic emission spectrometry (ICP/AES). These techniques require a high vacuum condition, or certain constraining operation pressures, or sample size requirement, otherwise the sample would be destroyed.

Calibration-free laser-induced breakdown spectroscopy (CF-LIBS) has been applied for precious multi-element analysis for various materials, including alloy metal [9], rock analogues, soil [10] and hydrogen isotopes retention at the plasma facing components (PFCs) in the international thermonuclear experimental reactor (ITER) tiles [11]. This approach is qualified to output accurate and continuous chemical compositions of samples without measuring standard curves through certified reference standard samples and internal standards [12–14].

In this work, such CF-LIBS technique was developed and used to analyze the trace of penetrated lithium and the distribution of matrix elements concentration in the corroded layer of CLF-1 specimen through depth profile analysis. The corroded layers display different component distribution after being exposed to the liquid lithium at 500 °C for 500 h. The surface of samples became coarse and porous. The results show that the depth profile includes the distinctive longitudinal variations within the corroded layer from interface to the inner substrate, which explains the quantitative elemental variations in the whole static corrosion process. This suggests that CF-LIBS has the capability to demonstrate the relative concentrations of the elements, which could be employed as an efficient approach for the quantitative analysis of lithium corrosion in fusion reactors.

2. Materials and Methods

2.1. Specimen Preparation

For this experiment, the substrates of CLF-1 with square size of $15 \times 15 \times 1$ mm^3 were initially prepared. The surfaces of the specimens were mechanically polished and cleaned by alcohol in ultrasonic machine (GT Sonic, Meizhou, China). The chemical composition of specimen is listed in Table 1.

Table 1. Chemical compositions of the China low-activation Ferritic steel (CLF-1) (wt.%).

Sample	Cr	Mn	C	W	V	Ta	N	Fe
CLF-1	8.5	0.5	0.11	1.5	0.26	0.1	0.03	Balance

2.2. Corrosion Experiment

The substrates were sealed into the Mo crucible in glove box (Vigor, Suzhou, China) with argon atmosphere and then heated at the constant temperature of 500 °C for 500 h. The static corrosion device (Self-developed by Insititue of mateirals, China Academy of Engineering Physics, Jiangyou, China) is made up of a vacuum chamber, pump, heating equipment, and control system. During the corrosion process, the samples were completely immersed in liquid lithium with argon atmosphere. After the experiment, the crucibles were taken out of the device. To avoid the intense reaction with water and to protect the corroded surface, the corroded specimens were cleaned by pure alcohol to remove the adhering lithium until the weight of the specimens remained constant. Through ultrasonic cleaning, all the adhering lithium solved and the whole corroded surface of the specimens were revealed.

2.3. Device and Setup of LIBS

The experimental LIBS set-up is shown in Figure 1. It is composed of laser, fiber-optic spectrometer, pulse generators, optical table, and a target holder. LIBS signals were obtained by using a Q-swiched Nd:YAG laser (LTB lasertechnik, Berlin, Germany) with 5 ns pulse width emitting at the fundamental wavelength of 1064 nm. The laser beam was focused onto the substrate surface by a 50-mm-focal-length bi-convex quartz lens.

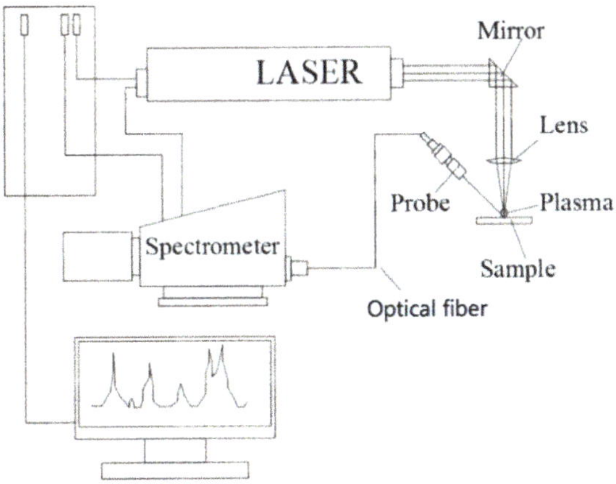

Figure 1. Diagram of Laser-Induced Breakdown Spectroscopy (LIBS) setup.

The analyzing laser beam was directed in a vertical direction to the sample surface which generated a crater with diameter of approximately 0.5 mm. To get high detection sensitivity, a calorimeter was used for adjusting the pulse energy at 70 mJ. After the laser-induced plasma has been set up, the light of the laser-induced plasma was gathered by a quartz collimating mirror (Avants UV-74) which was positioned on an optical table at 60° angle of incidence to the objective. Then, the emission signal was transferred by a long coupling fiber optic (core diameter 400 μm) into an echelle spectrometer with ICCD detector. A delay time was set at 1.0 μs and a gate time was set at 5.0 μs were employed for the plasma emission collection that were satisfied to lower background signals from continuous plasma shots and refrain from intense variations in plasma temperature in the process. In this work, the specimens were ablated by 30 consecutive shots on one spot, and the results were obtained by averaging 5 spectrums. The lines intensities become constant after 15th shoot.

3. Results

3.1. Microstructure

As shown in Figure 2, the surface of CLF-1 steel before corrosion was smooth and flat, while after exposure in lithium, the surface become rough and porous which would bring the disturbance of Matrix effect for LIBS measurement. Dark spots distributed on the surface have a high content of oxygen and carbon, which maybe formed by the difference of binary phase (Martensitic and prior austenitic) of the CLF-1. The results are shown in Appendix A Table A1 and Figure A1. The oxygen probably was brought by the impurity from the lithium and the cleaning process.

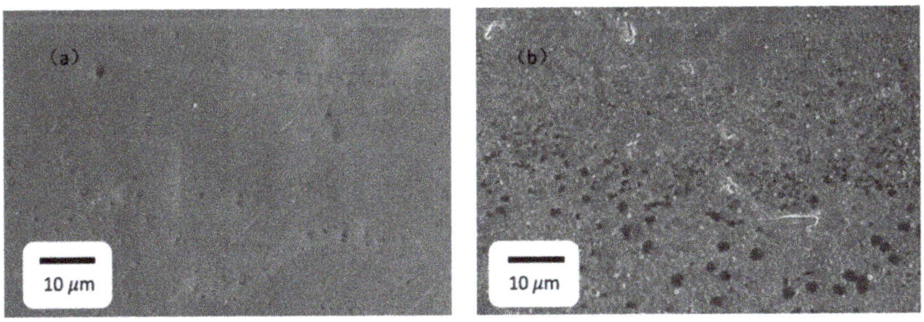

Figure 2. SEM images of CLF-1 (**a**) before liquid lithium corrosion; (**b**) after liquid lithium corrosion.

3.2. LIBS Spectra

In order to investigate the elemental dissolution and permeation helping us to understand the liquid lithium corrosion mechanism on structure material CLF-1, CF-LIBS was introduced to analyze the quantitative composition variation along the corroded layer. A sample LIBS spectrum is shown in Figure 3 with a low intensity signal-to-noise ratio. Combined with the National Institute of Standards and Technology (NIST) spectral database and the spectra [15], the influenced elements of the corroded CLF-1, i.e., Fe, Cr, Mn and Li could be clearly distinguished and detected by the characteristic spectrums. The emission lines of W and V with low content in CLF-1 are monitored as well.

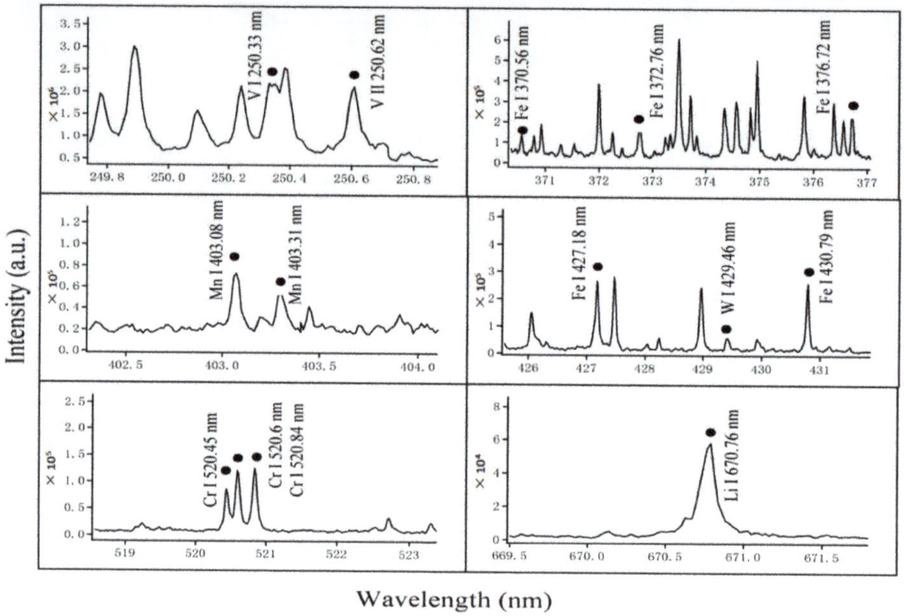

Figure 3. Typical LIBS spectra of CLF-1 steels after exposure to Li.

3.2.1. Evaluation of Local Thermodynamic Equilibrium and Matrix Effect

In this work, the inhomogeneous samples and the spectral responses of the spectrometer have to be taken into account, which may profoundly affect LIBS measurement. So, it is necessary to estimate the local thermodynamic equilibrium (LTE) conditions of laser-induced plasma and matrix effect to exclude the interference mentioned above [16]. When the LTE and matrix effect conditions are established and valid, the emission line is mostly identified by three parameters: the electron temperature, the electron density, and the elemental concentration [17].

Electron temperatures (Te) are estimated via Boltzmann plots of Fe atomic lines all selected from NIST database as isolated and with weak self-absorption and listed in Table 2. Taking the logarithm of the Boltzmann equation [18], we can represent each Fe I emission line as a point in Boltzmann plot demonstrated in Figure 4. Thus, the line intensity is figured as a Boltzmann plot where the slope of the lines is related to the electron temperature which fluctuates around 8000 K, as shown in Figure 5.

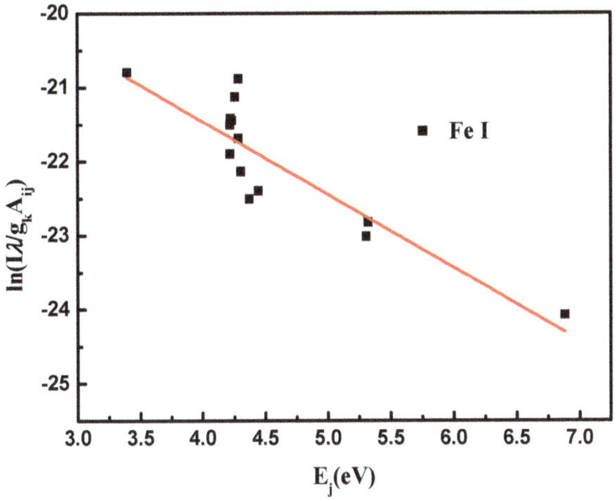

Figure 4. Boltzmann plot for Fe I lines.

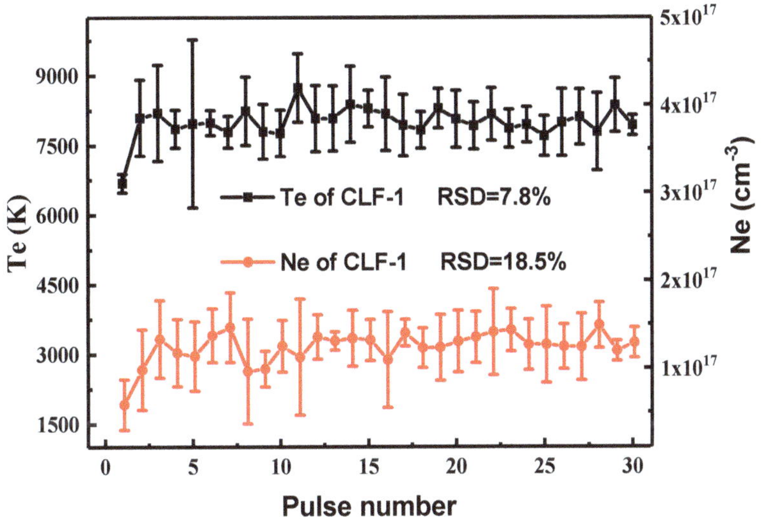

Figure 5. Depth profiles of the excitation temperature and the electron density on the CLF-1 steel surface.

Based on McWhirter's criterion [19], the necessary condition of LTE is determined by the electron density (*Ne*). Thus, to calculate electron density, Fe I line at 426.047 nm is chosen to measure *Ne* because of the good SNRs and no other overlapping spectral lines. The Stark broadening of Fe I 426.047 nm expressed by full width at half maximum (FWHM) is fitted by Lorentzian functions [20,21] as follows:

$$\Delta\lambda_{1/2} = 2\omega\cdot\left(N_e/10^{16}\right) \tag{1}$$

where $\Delta\lambda_{1/2}$ is the FWHM of the emission line, *Ne* is the electron number density of plasma, and ω is the electron width parameter. The *Ne* values in the range of 0.5–1.5 × 10^{17} cm^{-3} as function of LIBS shots is shown in Figure 5.

Based on the McWhirter criterion [17], the critical lower limit of *Ne* (cm^{-3}) for LTE can be described by:

$$Ne \geq 1.6 \times 10^{12} T^{\frac{1}{2}} (\Delta E)^3 \tag{2}$$

where T (K) refers to the plasma temperature, and ΔE (eV) the energy transition gap in this study equals to 2.18 eV [16]. The measured value of electron densities 1.5 × 10^{15} cm^{-3} is two orders lower than the electron densities measured by the stark broadening. Therefore, the value of *Ne* meets with the McWhirter conditions that establish the LTE conditions.

Figure 5 presents the curves with error bar of electron temperature and electron density for laser induced plasma. The relative standard deviation (RSD) of plasma temperature and electron density are 7.8% and 18.5%, respectively. For LIBS measurement, it is justified which the matrix effect couldn't be ignored [22].

Table 2. Fe I lines and parameters for Boltzmann plot method.

Species	Wavelength (nm)	A_{ij} (s^{-1})	g_k	E_i (eV)	E_j (eV)
Fe I	277.390	9.36 × 10^7	9	2.4534	6.8744
	281.329	3.42 × 10^7	11	0.9146	5.3204
	360.886	8.13 × 10^7	5	1.0111	4.4456
	363.146	5.17 × 10^7	9	0.9582	4.3714
	370.557	3.21 × 10^6	7	0.0516	3.3965
	372.762	2.24 × 10^7	5	0.9582	4.2833
	376.719	6.39 × 10^7	3	1.0111	4.3013
	378.788	1.29 × 10^7	5	1.0111	4.2833
	379.500	1.15 × 10^7	7	0.9901	4.2562
	425.079	1.02 × 10^7	7	1.5574	4.4733
	426.047	3.99 × 10^7	11	2.3992	5.3085
	427.176	2.28 × 10^7	11	1.4849	4.3865
	430.790	3.38 × 10^7	9	1.5574	4.4346
	438.354	5.00 × 10^7	11	1.4849	4.3125
	440.475	2.75 × 10^7	9	1.5574	4.3714

3.2.2. Chemical Depth Profile of CF-LIBS Results

When the LTE condition is met, the CF-LIBS method advises that the line integral intensity corresponding to elemental concentration can be expressed as [14]:

$$\overline{I_{\lambda}^{ki}} = FC_sA_{ki}\frac{g_k e^{-(E_k/K_BT)}}{Q_s(T)} \tag{3}$$

where $\overline{I_{\lambda}^{ki}}$ is the measured integral line intensity, $Q_s(T)$ is the partition function, C_s is the concentration of the emitting atomic element, A_{ki} is the transition probability, g_k and E_k are the energy and degeneracy of the level retrieved from the NIST database and F is the parameter which determines the optical efficiency of the acquisition system, like the plasma energy and density. In the whole experimental process, it is important for the stability of the F constant to remain the within the experimental parameters fixed in a certain measurement range, such as laser energy, focus, optical path, and so on.

The logarithm of Equation (3) yields a mathematical method for the determination of electron temperature. Thus, the line intensity can be graphically represented as a Boltzmann plot where the slope and intercept (q_s) of the straight line is correlated with the electron temperature and the elements concentration, respectively. In the LTE theory, once the electron temperature is determined, the concentration of the element can be measured by the calculation of the spectral intensity. F could be identified by the unitary sum of the element concentration C_s [14].

$$\sum_S C_s = \frac{1}{F}\sum_s Q_s(T)e^{q_s} = 1 \tag{4}$$

The concentration of all the elements in specimens can be calculated from:

$$C_s = \frac{Q_s(T)}{F} e^{q_s} \tag{5}$$

To optimize the results and accuracy, the selection of the spectral line has to consider other several parameters. First, the neutral and singly ionized lines with excited lower state are preferred over the ground state to avoid the self-absorption. Second, we calculate the mean value of spectrum by averaging the line intensities of each laser pulse respectively for five times. The selected spectral lines, listed in Table 3, have good SNRs and weak self-absorption, indicating that those typical matrix elements as well as Li have the good qualification to be detected. In this experiment, a number of 15 spectra are consecutively recorded to demonstrate the quantitative elemental content with their intensities. The concentration curve variations as a function with pulse number may reflect the mechanism of the elemental transaction in the depth profile during the corrosion process, shown in Figure 6.

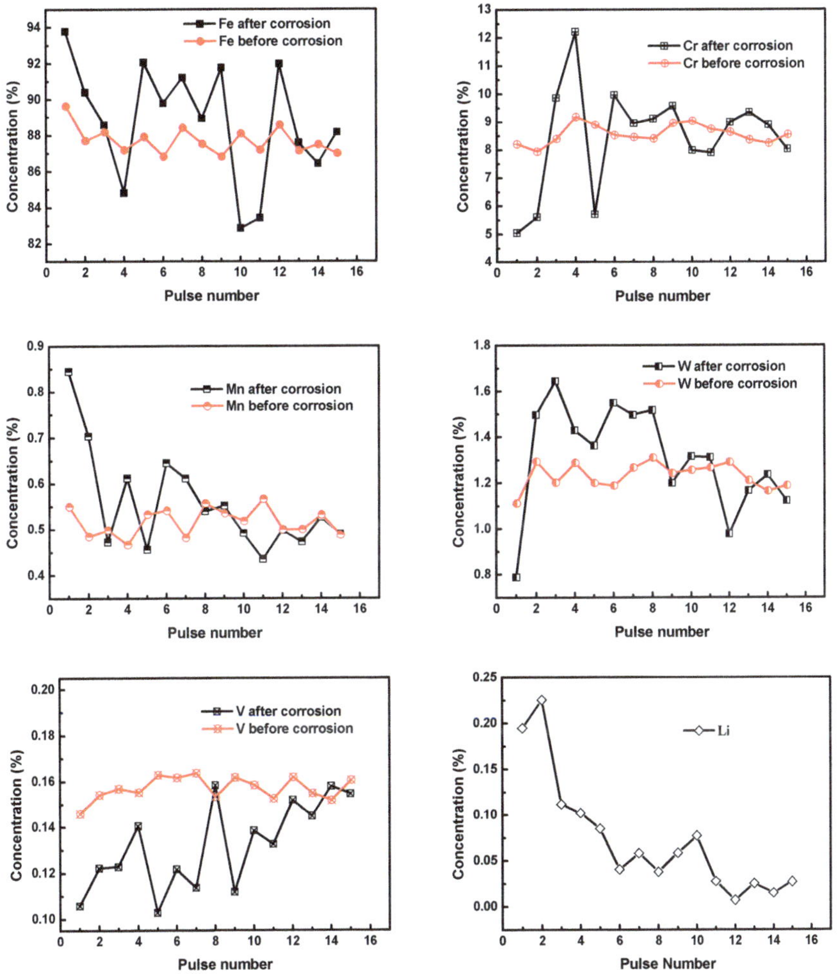

Figure 6. Depth profiles of elemental quantitative concentration normalized by CF-LIBS method for CLF-1 before and after exposure to liquid lithium. The pulse numbers approximately represent the value of depth.

Table 3. Selected emission lines for depth profile analysis.

Species	Wavelength (nm)	A_{ij} (s^{-1})	g_k	E_i (eV)	E_j (eV)
V I	250.330	4.40×10^7	4	0.0000	4.9513
	306.638	2.10×10^8	12	0.0685	3.9597
	318.398	2.50×10^8	5	0.0000	3.1758
	438.472	1.10×10^8	8	0.0000	3.0751
V II	250.622	9.72×10^7	9	1.0962	6.0418
	309.310	2.00×10^8	13	0.3921	4.3994
W I	386.798	4.60×10^6	3	0.0000	3.8807
	429.460	1.24×10^7	3	0.0000	3.2633
Mn I	279.480	3.70×10^8	58	0.0000	4.4349
	403.076	1.65×10^7	6	0.0000	3.0751
	403.307	1.58×10^7	4	0.0000	3.0733
	403.450	1.58×10^7	4	0.0000	3.0722
	404.141	7.87×10^7	10	2.1142	5.1812
Mn II	257.615	2.80×10^8	9	0.0000	4.8114
	294.928	1.96×10^8	7	1.1745	5.3772
	348.870	2.11×10^7	3	1.8475	5.4004
Cr I	425.440	3.15×10^7	9	0.0000	2.9134
	427.480	3.07×10^7	7	0.0000	2.8995
	428.970	3.16×10^7	5	0.0000	2.8894
	520.450	5.09×10^7	3	0.9414	3.3230
	520.602	5.14×10^7	5	0.9414	3.3223
	520.841	5.06×10^7	7	0.9414	3.3212
Cr II	284.980	9.20×10^7	8	1.5100	5.8622
	286.090	6.90×10^7	4	1.4800	5.8245
	286.263	6.30×10^7	8	1.5300	5.8683
	286.671	1.20×10^8	4	1.4918	5.8154
	311.865	1.70×10^8	4	2.4211	6.3956
	312.041	1.50×10^8	6	2.4339	6.4061
Li I	610.354	5.71×10^7	4	1.8478	3.8786
	670.776	1.47×10^8	4	0.0000	1.8478
Fe I	360.886	8.13×10^7	5	1.0111	4.4456
	363.146	5.17×10^7	9	0.9582	4.3714
	370.557	3.21×10^6	7	0.0516	3.3965
	372.762	2.24×10^7	5	0.9582	4.2833
	376.719	6.39×10^7	3	1.0111	4.3013
	378.788	1.29×10^7	5	1.0111	4.2833
	379.500	1.15×10^7	7	0.9901	4.2562
	400.524	2.04×10^7	5	1.5574	4.6520
	404.581	8.62×10^7	9	1.4849	4.5485
	407.174	7.64×10^7	5	1.6079	4.6520
	425.079	1.02×10^7	7	1.5574	4.4733
	426.047	3.99×10^7	11	2.3992	5.3085
	427.176	2.28×10^7	11	1.4849	4.3865
	430.790	3.38×10^7	9	1.5574	4.4346
	438.354	5.00×10^7	11	1.4849	4.3125
	440.475	2.75×10^7	9	1.5574	4.3714
Fe II	261.187	1.20×10^8	8	0.0477	4.7932
	261.382	2.12×10^8	2	0.1069	4.8489
	262.166	5.60×10^7	2	0.1211	4.8489
	262.566	3.52×10^7	10	0.0477	4.7683
	262.829	8.74×10^7	4	0.1211	4.8370
	263.104	8.16×10^7	6	0.1069	4.8178
	266.466	1.91×10^8	10	3.3866	8.0381
	268.475	1.57×10^8	10	3.8143	8.4310
	298.554	2.39×10^7	4	1.7239	5.8755

In the liquid metal corrosion mechanism, the corrosion medium lithium permeates into the crystal boundary phase by thermal diffusion and mass transfer with the substrate compositions [23]. Figure 6 shows that the quantitative content curve of Li penetrated into the substrate from the intersurface to the inner matrix. The concentration of Li contents decreases dramatically from 0.2 wt.% to the almost zero. At 12th shot, the concentration of Li is close to zero. The area from the first shot to the 12th shot represents the permeated depth of lithium, and we consider this area as a corrosion layer.

Since Fe and Cr are the main components, their relative concentrations would affect each other. The concentration of Fe decreases from 94% to 84% in the corrosion layer. The concentration fluctuations is due to the binary phase (martensite and austenite) of CLF-1, where lots of Fe-C compounds form segregation in high temperature corrosion [23–26]. In the case of Cr, it is clearly demonstrated that the concentration of Cr fluctuates from 5% to 12% at the first six shots, then tend to be stable at 8.5%. During the corrosion process, Cr present in the crystal boundary in a form of carbon compounds ($M_{23}C_6$, M = Cr, V, W) which is exchanged by Li in the diffusive and dissovled process [27,28]. In the meantime, the spread speed of Cr in the crystal grain is relatively slow causing the reduction of Cr in the grain boundary that cannot be replenished in time. In comparison, the relative concentration of Cr dissolution is higher than Fe which may be attributed to higher diffusion coefficient of Cr to Fe in liquid Li [29]. In the case of Mn, the diffusion rate of Mn from the inner layer to the interface is quick in exposure to liquid lithium. When Mn in the interface dissolved into the liquid lithium, Mn transfer from the interior to the interface made the concentration of Mn of up to 0.8%. The concentration of V increases from 0.11% to 0.16% along the corrosion layer. V is sensitive to the exchange of non-metals like nitrogen that induce the reduction of concentration near the surface regions [30,31]. The calculated concentration of V before and after the corrosion was lower than the concentration of 0.26%, which could be caused by the relatively low content in the substrate. Further, for the element W, because of the solution of W in liquid lithium is low [31], the concentration change of W is not evident, and decreases from 1.6% to 1.2%.

4. Conclusions

The structural material CLF-1 has been exposed to liquid lithium at 500 °C for 500 h. The surface of CLF-1 became coarse and trace lithium was found in the surface. The calibration-free LIBS method was introduced to perform quantitative elemental composition depth profiles analysis for CLF-1 exposed to liquid lithium. The results demonstrate the ability of CF-LIBS for in-depth chemical composition analysis of the corrosive specimens and obtained different element distributions in the corrosion layer, including trace element and corrosion medium Li. It should be noted that this method would produce scattered values because of highly inhomogeneous corroded samples and binary-phase which made the sample composition segregated.

From the surface to the inner layer, the concentration of Fe, Mn, W, and Li decreased from 95% to 83%, 0.85% to 0.45%, 1.6% to 1.1%, and 0.2% to almost 0, respectively. Cr increased from 0.11% to 0.16%. V fluctuated between 5% and 12%. The elemental variation in the depth profile represents a considerable index to explain the corrosion resistance of steels within liquid lithium. This study shows that the CF-LIBS measurement represents a useful procedure for the quantitative measurement of the elemental variation of corroded structural materials in the fusion blankets. It provides a fast, remote, real-time, and convenient way to monitor the compatibility of the liquid metal blankets for fusion reactor applications.

Author Contributions: Conceptualization, Z.C. and Y.A.; validation, C.C., Y.A. and Y.L.; formal analysis, Y.A.; investigation, Z.C.; resources, C.C.; writing—original draft preparation, Z.C.; writing—review and editing, Z.C. and X.W.; visualization, Z.C.; supervision, C.C.; project administration, Y.A.; funding acquisition, C.C. and Y.A. All authors have read and agreed to the published version of the manuscript.

Funding: This research is funded by the financial support from National Key Research and Development Program of China (2017YFE0301506 and 2015GB109004).

Acknowledgments: The authors would like to acknowledge the support of CLF-1 samples of China Academy of Science and Southwestern Institute of Physics, for the CLF-1. Deeply appreciate the help of Tao Xu and Gou for the LIBS analysis.

Conflicts of Interest: The authors declare no conflict of interest.

Appendix A

Table A1. Composition of the CLF-1 at the light spot and the dark spot exposed to Li analyzed by EDS.

Element (wt.%)	O	C	Cr	Fe	W	Mn	V
Light spot	5.58	3.5	8.2	80.67	1.1	0.40	0.55
Dark spot	10.90	6.36	6.95	71.99	1.6	-	2.2

To explore the dark spots in the Figure 2, we use the XRD pattern that Li compounds (Li_2CO_3) were found on the surface which maybe the dark spot on the surface observed on the SEM. In this experiment, the cleaning medium is alcohol to solve the lithium adhered to the surface not intensely react like flowing water. This phenomenon explains why the intensity of lithium spectral line in the surface are so high for the first few impulses.

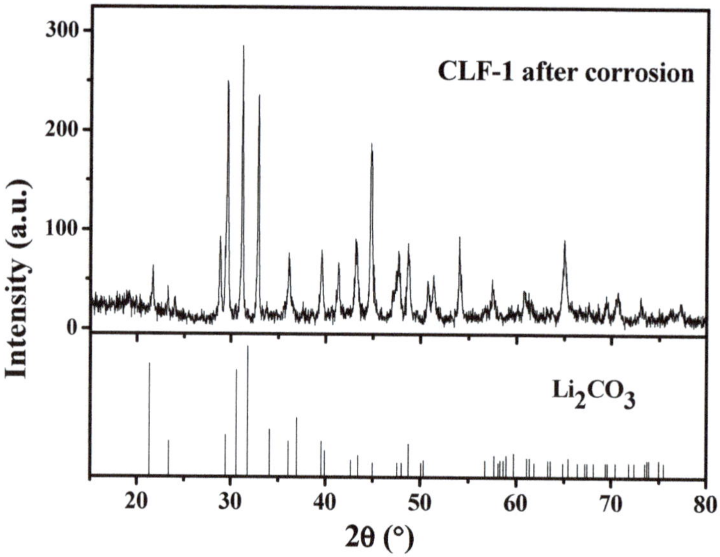

Figure A1. XRD pattern of the CLF-1 steel after corrosion.

References

1. Mansfield, D.K.; Hill, K.W.; Strachan, J.D. Enhancement of Tokamak Fusion Test Reactor performance by lithium conditioning. *Phys. Plasmas* **1996**, *3*, 1892–1897. [CrossRef]
2. Wong, C.P.C.; Salavy, J.F.; Kim, Y.; Kirillov, I.; Rajendra Kumar, E.; Morley, N.B.; Tanaka, S.; Wu, Y.C. Overview of liquid metal TBM concepts and programs. *Fusion Eng. Des.* **2008**, *83*, 850–857. [CrossRef]
3. Mansfield, D.K.; Johnson, D.W.; Grek, B.; Kugel, H.W.; Bell, M.G.; Bell, R.E.; Budny, R.V.; Busha, C.E.; Fredrickson, E.D.; Hill, K.W. Observations concerning the injection of a lithium aerosol into the edge of TFTR discharges. *Nucl. Fusion* **2000**, *41*, 1823–1834. [CrossRef]
4. Yamaki, E.; Ginestar, K.; Martinelli, L. Dissolution mechanism of 316L in Lead-bismuth eutectic at 500 °C. *Corros. Sci.* **2011**, *53*, 3075–3085. [CrossRef]
5. Muroga, T.; Gasparotto, M.; Zinkle, S.J. Overview of materials research for fusion reactors. *Fusion Eng. Des.* **2002**, *61*, 13–25. [CrossRef]

6. Zhang, M.; Team, F.D.S. Corrosion experiment for CLAM and SS316L in liquid LiPb loop of China. *Ann. Nucl. Energy* **2015**, *80*, 203–206. [CrossRef]
7. Wang, P.; Chen, J.; Fu, H.; Liu, S.; Li, X.; Xu, Z. Technical issues for the fabrication of a CN-HCCB-TBM based on RAFM steel CLF-1. *Plasma Sci. Technol.* **2013**, *15*, 133–136. [CrossRef]
8. Oriani, R.A. Intergranular corrosion of steels and alloys. *Mater. Sci. Eng.* **1985**, *73*, 219. [CrossRef]
9. Burakov, V.S.; Kiris, V.V.; Naumenkov, P.A.; Raikov, S.N. Calibration-free laser spectral analysis of glasses and copper alloys. *J. Appl. Spectrosc.* **2004**, *71*, 740–746. [CrossRef]
10. Colao, F.; Fantoni, R.; Lazic, V.; Paolini, A.; Fabbri, F.; Ori, G.G.; Marinangeli, L.; Baliva, A. Investigation of LIBS feasibility for in situ planetary exploration: An analysis on Martian rock analogues. *Planet. Space Sci.* **2004**, *52*, 117–123. [CrossRef]
11. Fantoni, R.; Almaviva, S.; Caneve, L.; Colao, F.; Popov, A.M.; Maddaluno, G. Development of calibration-free laser-induced-breakdown-spectroscopy based techniques for deposited layers diagnostics on ITER-like tiles. *Spectrochim. Acta B* **2013**, *87*, 153–160. [CrossRef]
12. Thakur, S.N.; Singh, J.P. Fundamentals of Laser Induced Breakdown Spectroscopy. In *Laser-Induced Breakdown Spectroscopy*; Elsevier: Amsterdam, The Netherland, 2007; Chapter 1; pp. 3–21.
13. Pandhija, S.; Rai, N.K.; Rai, A.K.; Thakur, S.N. Contaminant concentration in environmental samples using LIBS and CF-LIBS. *Appl. Phys. B* **2010**, *98*, 231–241. [CrossRef]
14. Ciucci, A.; Corsi, M.; Palleschi, V.; Rastelli, S.; Salvetti, A.; Tognoni, E. New procedure for quantitative elemental analysis by laser-induced plasma spectroscopy. *Appl. Spectrosc.* **1999**, *53*, 960–964. [CrossRef]
15. NIST Data Base. Available online: http://physics.nist.gov/PhysRefData/ASD/lines_form.html (accessed on 20 December 2019).
16. Ma, Q.; Motto-Ros, V.; Lei, W.; Boueri, M.; Zheng, L.; Zeng, H.; Bar-Matthews, M.; Ayalon, A.; Panczer, G.; Yu, J. Multi-elemental mapping of a speleothem using laser-induced breakdown spectroscopy. *Spectrochim. Acta B* **2010**, *65*, 707–714. [CrossRef]
17. Ng, C.W.; Ho, W.F.; Cheung, N.H. Spectrochemical analysis of liquids using laser-induced plasma emissions: Effects of laser wavelength on plasma properties. *Appl. Spectrosc.* **1997**, *51*, 967–983. [CrossRef]
18. Cristoforetti, G.; DeGiacomo, A.; Dell'Aglio, M.; Legnaioli, S.; Tognoni, E.; Palleschi, V.; Omenetto, N. Local thermodynamic equilibrium in laser-induced breakdown spectroscopy: Beyond the McWhirter criterion. *Spectrochim. Acta B* **2010**, *65*, 86–95. [CrossRef]
19. Nishijima, D.; Doerner, R.P.; Hollmann, E.M.; Miyamoto, M. Laser-induced breakdown spectroscopy thickness measurements of films thinner than ablation rate. *Spectrochim. Acta B* **2017**, *136*, 34–38. [CrossRef]
20. Xu, T.; Zhang, Y.; Zhang, M.; He, Y.; Yu, Q.; Duan, Y. Temporal-resolved characterization of laser-induced plasma for spectrochemical analysis of gas shales. *Spectrochim. Acta B* **2016**, *121*, 28–37. [CrossRef]
21. Sabsabi, M.; Cielo, P. Quantitative analysis of aluminum alloys by laser-induced breakdown spectroscopy and plasma characterization. *Appl. Spectrosc.* **1995**, *49*, 499–507. [CrossRef]
22. Xu, T.; Liu, J.; Shi, Q.; He, Y.; Niu, G.; Duan, Y. Multi-elemental surface mapping and analysis of carbonaceous shale by laser-induced breakdown spectroscopy. *Spectrochim. Acta B* **2016**, *115*, 31–39. [CrossRef]
23. Chopra, O.K.; Smith, D.L. Effects of lithium environment on the fatigue properties of ferritic and austenitic steels. *J. Nucl. Mater.* **1981**, *103*, 651–656. [CrossRef]
24. Li, Y.; Ke, C.; Liu, X.; Gou, F.; Duan, X.; Zhao, Y. Compositional depth profiles of the type 316 stainless steel undergone the corrosion in liquid lithium using laser-induced breakdown spectroscopy. *J. Nucl. Mater.* **2017**, *497*, 1–9. [CrossRef]
25. Xu, Q.; Kondo, M.; Nagasaka, T.; Muroga, T.; Nagura, M.; Suzuki, A. Compatibility of Reduced Activation Ferritic Steels with Liquid Lithium. *Fusion Sci. Technol.* **2007**, *52*, 609–612. [CrossRef]
26. Leavenworth, H.W.; Cleary, R.E. The solubility of Ni, Cr, Fe, Ti and Mo in liquid lithium. *Acta Met.* **1961**, *9*, 519–520. [CrossRef]
27. Eremenko, V.N.; Natanzon, Y.V. Kinetics of external dissolution of metals in metallic melts (Review). *Sov. Powder Met. Met. Ceram.* **1970**, *9*, 645–658. [CrossRef]
28. Chakraborty, P.; Kain, V.; Pradhan, P.K.; Fotedar, R.K.; Krishnamurthy, N.; Dey, G.K. Corrosion of Indian RAFMS in Pb–17Li in a rotating disc corrosion test facility at 773 K. *Fusion Eng. Des.* **2015**, *100*, 181–189. [CrossRef]

29. Chopra, O.K.; Smith, D.L. Compatibility of ferritic steels in forced circulation lithium and Pb-17Li systems. *J. Nucl. Mater.* **1988**, *155*, 715–721. [CrossRef]
30. Atchutuni, S.S.; Agravat, H.S.; Chauhan, J.P. Corrosion behavior of IN-RAFM steel with stagnant lead-lithium at 550 °C up to 9000 h. *Fusion Eng. Des.* **2018**, *135*, 102–109. [CrossRef]
31. Votinov, S.N.; Kolotushkin, V.P.; Lyublinskii, I.E.; Vertkov, A.V.; Nikulin, S.A.; Turilina, V.Y. Corrosion resistance of vanadium alloys clad by a ferrtic corrosion-resistant steel in liquid-metal heat-transfer agents. *Russ. Metall.* **2009**, *1*, 93–98. [CrossRef]

materials

MDPI

Article

Effect of Residual $CaSO_4$ in Clinker on Properties of High Belite Sulfoaluminate Cement Based on Solid Wastes

Dunlei Su [1], Qiuyi Li [2,*], Yuanxin Guo [2,*], Gongbing Yue [2] and Liang Wang [2]

1 School of Civil Engineering, Qingdao University of Technology, Qingdao 266033, China; sudunlei@163.com
2 School of Architectural Engineering, Qingdao Agricultural University, Qingdao 266109, China; yuegongbing@163.com (G.Y.); jiangongwl_2019@163.com (L.W.)
* Correspondence: lqyyxn@163.com (Q.L.); guoyuanxin2019@163.com (Y.G.)

Received: 6 November 2019; Accepted: 13 January 2020; Published: 16 January 2020

Abstract: The high belite sulfoaluminate cement (HBSAC) containing $CaSO_4$, and without $CaSO_4$, based on solid wastes were successfully prepared from petroleum coke desulfurization slag (PCDS), fly ash (FA), carbide slag (CS), and bauxite (BX). The mineral composition of clinkers after different calcination history were investigated by X-ray fluorescence (XRF), X-ray diffraction (XRD)/Quantitative X-ray diffraction (QXRD), and scanning electron microscopy (SEM), so as to determine the calcination temperatures. The difference between residual $CaSO_4$ and dihydrate gypsum (DG) and the optimal content of residual $CaSO_4$ were discussed by studying the properties of HBSAC. The results revealed that the residual $CaSO_4$ in clinker could replace DG to participate in hydration, and showed some advantages in strength and early hydration heat, but meanwhile increased the water requirement of normal consistency and hydration heat at 72 h, and prolonged the setting time. With the increase of residual $CaSO_4$ content in clinker, the lower limit temperature of clinker formation gradually increased, and the crystal size of clinker minerals became finer and the boundary between crystals became more blurred. However, the optimal calcination temperature (1300 °C) of HBSAC clinker did not change. Considering the effect of residual $CaSO_4$ content on the water requirement of normal consistency, setting time, hydration heat, strength, and hydration products, the optimal design content of residual $CaSO_4$ in HBSAC clinker based on solid wastes, such as PCDS and FA, was 15%.

Keywords: residual $CaSO_4$; solid wastes; high belite sulfoaluminate cement; petroleum coke desulfurization slag; $CaSO_4$ type; $CaSO_4$ content; cement properties

1. Introduction

As an inorganic nonmetallic material with obvious advantages, such as high early strength, good later strength development, low energy consumption, and CO_2 emission, high belite sulfoaluminate cement (HBSAC) has become a hot spot in the field of sulfoaluminate cement (SAC). The mineral components of this cement clinker are mainly $C_4A_3\overline{S}$ ($3CaO{\cdot}3Al_2O_3{\cdot}CaSO_4$) and β-C_2S ($2CaO{\cdot}SiO_2$). The raw materials for production are usually limestone, clay, bauxite, and gypsum [1,2]. However, with the long-term and large-scale exploitation of natural resources, the above-mentioned raw materials are becoming more and more scarce, and all kinds of industrial solid wastes are beginning to be used as alternative raw materials, including calcium solid waste materials such as marble sludge waste [3], steel slag [4,5], lithium slag [6] and municipal solid incineration wastes [7], sulfur solid waste materials such as phosphogypsum [8,9] and desulfurization gypsum [10,11], and aluminum–silicon solid waste materials such as fly ash (FA) [11–16], coal gangue [17], red mud [10,18], aluminum anodizing sludge [19], and tailings [20]. $C_4A_3\overline{S}$ is an indispensable component in the hydration of HBSAC, and the main chemical reactions in the hydration process are shown in Equations (1) and

(2). $C_4A_3\overline{S}$ can be hydrated to form AFt ($C_3A{\cdot}3C\overline{S}{\cdot}32H$) under the condition of sufficient $CaSO_4$, while $C_4A_3\overline{S}$ can be hydrated to form AFm ($C_3A{\cdot}C\overline{S}{\cdot}12H$) under the condition of insufficient $CaSO_4$. Compared with AFm, AFt has higher strength and plays a more important role in the development of cement strength, so it seems to be a reasonable choice to have some residual $CaSO_4$ in clinker, which can not only reduce the amount of natural gypsum, but also greatly improve the utilization rate of sulfur solid waste materials.

$$C_4A_3\overline{S} + 2C\overline{S}H_2 + 34H \rightarrow C_3A{\cdot}3C\overline{S}{\cdot}32H\,(AFt) + 2AH_3(gel). \quad (1)$$

$$C_4A_3\overline{S} + 18H \rightarrow C_3A{\cdot}C\overline{S}{\cdot}12H\,(AFm) + 2AH_3(gel). \quad (2)$$

The preparation of HBSAC clinker containing $CaSO_4$ from solid wastes has been studied before. Chen et al. [14] successfully prepared HBSAC clinker with mineral content of about 43% $C_4A_3\overline{S}$, 42% C_2S, 8% C_4AF ($4CaO{\cdot}Al_2O_3{\cdot}Fe_2O_3$), and 3% $CaSO_4$ from industrial wastes such as FA, flue gas desulfurization sludge, and fluidized bed ash. Lv et al. [21] prepared HBSAC clinker with mineral content of about 20% $C_4A_3\overline{S}$, 56% C_2S, 17% C_2AF ($2CaO{\cdot}Al_2O_3{\cdot}Fe_2O_3$), and 3.5% $CaSO_4$ by using limestone, bauxite (BX), and ashes from circulating fluidized bed combustion. However, the residual $CaSO_4$ content in both clinkers was low, and it was often necessary to admix a certain amount of natural gypsum with the clinker to make the cement achieve better performance, which could not save the natural gypsum resources to a large extent. For example, HBSAC clinker prepared by Chen et al. needed 20% natural dihydrate gypsum (DG) to achieve stable exothermic rate, but the cement pastes exhibited expansion cracking under this condition, which needed 1% citric acid retarder to eliminate. When 20% natural DG and 1% citric acid retarder were added, the compressive strength of cement mortar prepared according to ASTM C 109 standard could reach 15, 24, 29, and 45 MPa at 1, 3, 7, and 28 days, respectively. As for HBSAC clinker prepared by Lv et al., the compressive strength of cement pastes under the condition of water:cement ratio of 0.3 could reach 25 and 45 MPa at 3 and 28 days, respectively. When 10% natural DG was added, the compressive strength of cement pastes could reach the maximum value, which was 35 and 80 MPa at 3 and 28 days, respectively.

Tangshan Polar Bear Building Materials Co., Ltd. [11] has produced HBSAC clinker with mineral composition of 20%–35% $C_4A_3\overline{S}$, 37%–47% C_2S, 3%–9% C_4AF, 0.5%–4.6% f-CaO, and 14%–26.3% $CaSO_4$ by using FA, desulfurization gypsum, and other natural raw materials. This clinker could be directly used as cement, and showed many advantages, such as fast setting and hardening, early strength and high strength, low shrinkage, impermeability, frost resistance, corrosion resistance, and so on. Nowadays, the relevant production technology of this clinker has been relatively perfect, and good social benefits have been achieved in the engineering applications of Qingdao Airport and oilfields of Dongying. Taking HBSAC clinker with mineral composition of 28.6% $C_4A_3\overline{S}$, 39.6% C_2S, 4.6% C_4AF, 1.6% f-CaO, and 13.6% $CaSO_4$ as an example, its water requirement of normal consistency was 31.3%, and its initial and final setting time were 9 and 13 min, respectively. The compressive strength of cement mortar prepared under the condition of water:cement ratio of 0.5 could reach over 20 MPa at 4 h, 30 MPa at 1 day, and 45 MPa at 28 days. At the same curing age, the bending strength could reach 4.3, 5.1, and 7.6 MPa, respectively. The cumulative heat released from hydration at 3 days was approximately 280 J/g. When mixed with anhydrite in cement clinker, the anhydrite content had different effect on various cement properties. With the increase of anhydrite content, the setting time did not change much and the water requirement of normal consistency decreased. In terms of compressive strength, the early compressive strength increased and the later strength decreased under the condition of 10% anhydrite, while the early compressive strength decreased and the later strength increased under the condition of 20% anhydrite.

Shen et al. [8] used phosphogypsum, FA, and other natural raw materials to prepare a series of HBSAC clinkers with residual $CaSO_4$ content ranging from 0% to 20%. Moreover, the other mineral composition of these clinkers were approximately 25%–40% $C_4A_3\overline{S}$, 30%–65% C_2S, and 10% C_4AF. The residual $CaSO_4$ in clinker could achieve similar effect to that of additional natural gypsum, and was even better than additional natural gypsum in cement properties. Meanwhile, the residual $CaSO_4$ content in clinker also had different effect on cement properties. In the aspect of setting time, there was little difference when the residual $CaSO_4$ content was lower than 10%, and there was an obvious growth trend when the residual $CaSO_4$ content was higher than 10%. The strength increased gradually with the increase of residual $CaSO_4$ content, while the expansion rate did not change obviously, and the strength increased most significantly when the residual $CaSO_4$ was 10%. For HBSAC clinker with mineral composition of 35.0% $C_4A_3\overline{S}$, 32.3% C_2S, 8.4% C_4AF, and 16.6% $CaSO_4$, when the water:cement ratio was 0.5, the initial and final setting time of cement paste were 130 and 220 min, the compressive strength of cement paste could reach 25, 30, 36, and 38 MPa at 1, 3, 7, and 28 days, respectively. At the same time, the cumulative heat released from hydration at 1 day reached 210 J/g. The expansion rate of cement mortar was basically stable at 1.5×10^{-4} according to the "Standard test method for drying shrinkage of mortar" (JC/T 603-2004, China).

It was not difficult to find that remarkable achievements had been made in the preparation of HBSAC clinker containing $CaSO_4$ from solid wastes, but there were still some problems to be solved urgently. Firstly, the method of improving cement properties by adding gypsum into HBSAC clinker with low $CaSO_4$ content not only wasted natural resources, but also brought more uncertainty to cement hydration because the effect of adding gypsum was not completely equal to the residual $CaSO_4$ in clinker. Secondly, even though HBSAC clinker with high $CaSO_4$ content could show excellent properties without adding gypsum, the properties of HBSAC clinker prepared with different solid wastes were also quite different, such as HBSAC clinkers prepared by Tangshan Polar Bear Co., Ltd. (Tangshan, China) and Shen et al. Petroleum coke desulfurization slag (PCDS) is the residue of sulfur-containing petroleum coke and desulfurizer (usually limestone) calcined at high temperature in a circulating fluidized bed boiler, which is composed of CaO and $CaSO_4$. It can not only replace the limestone which produces a lot of CO_2 and the natural gypsum which is in short supply, providing calcium and sulfur elements needed for the formation of clinker minerals such as $C_4A_3\overline{S}$ and C_2S, but also provide raw materials for residual $CaSO_4$ in clinker. In this study, PCDS, FA, carbide slag (CS), and BX were used together to prepare HBSAC clinker, including some containing different $CaSO_4$ contents and some without $CaSO_4$. At the same time, the effect of residual $CaSO_4$ content on calcination temperature of clinkers was discussed. Based on the study of cement properties, the effect of residual $CaSO_4$ in clinker and DG were compared, and the optimal content of residual $CaSO_4$ was determined, which would provide theoretical basis for the reduction and absorption of solid wastes, such as petroleum coke desulfurization slag, and the proportional optimum design of HBSAC based on solid wastes.

2. Experiment

2.1. Raw Materials

In the experiment, PCDS was taken from the Sinopec Qingdao Refining and Chemical Co., Ltd. (Qingdao, China), FA (Class I) from the Qingdao Municipal Concrete Industry Co., Ltd. (Qingdao, China), CS from the Qingdao Qingxin Building Materials Co., Ltd. (Qingdao, China), BX from the Gongyi Wanying Environmental Protection Materials Co., Ltd. (Gongyi, China), and DG from the Sinopharm Chemical Regent Co., Ltd. (Shanghai, China). The main chemical compositions of raw materials are listed in Table 1, and the main mineral compositions of PCDS, FA, CS, and BX are shown in Figure 1.

Table 1. Chemical composition of the raw materials, wt.%.

Raw Material	CaO	Al_2O_3	SiO_2	Fe_2O_3	SO_3	MgO	TiO_2	LOI	Σ
PCDS	52.93	0.96	4.56	1.13	30.12	2.03	0.00	7.48	99.21
FA	7.86	27.45	52.56	4.24	1.26	1.12	0.99	1.73	97.21
CS	66.02	1.47	4.61	0.68	1.97	0.25	0.00	24.62	99.62
BX	0.51	64.07	14.53	0.88	0.00	15.38	2.56	1.03	98.96
DG	32.56	/	/	/	46.51	/	/	20.92	99.99

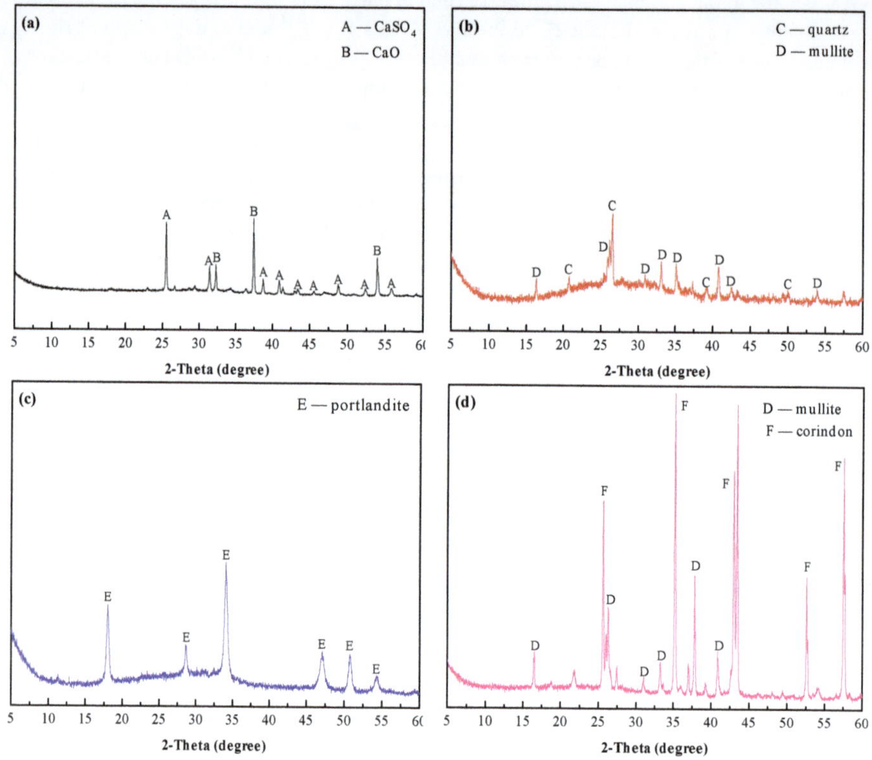

Figure 1. XRD patterns of raw materials: (**a**) PCDS; (**b**) FA; (**c**) CS; (**d**) BX.

2.2. Materials Proportioning

Based on the ternary mineral system of SAC clinker containing $C_4A_3\overline{S}$, C_2S and C_4AF, $CaSO_4$ was introduced into the mineral system of new clinker. Meanwhile, considering that the C_2S content in HBSAC clinker is generally more than 40%, the mineral composition of the clinkers and mixing design of raw materials are shown in Table 2. The Series GF-HBSAC-X cement was used to study the effect of $CaSO_4$ type on the properties of HBSAC based on solid wastes, and the Series GF-HBSAC-Y cement was used to study the effect of residual $CaSO_4$ content on the properties of HBSAC based on solid wastes.

In the process of developing a mixing design, the design principles are as follows: (1) The mineral composition of clinker is preliminarily set up, assuming that the following three reactions occur during calcination: $4CaO + Al_2O_3 + Fe_2O_3 \rightarrow C_4AF$, $3CaO + 3Al_2O_3 + CaSO_4 \rightarrow C_4A_3\overline{S}$, $2CaO + SiO_2 \rightarrow C_2S$, and the amount of raw materials was deduced according to Table 1. (2) As a raw material of calcium and sulfur, PCDS contains high $CaSO_4$ content. $CaSO_4$ is decomposed by heat in high temperature reaction, which will affect the amount of PCDS. After being calcined to a constant weight at 950 °C, the mass loss of PCDS calcined at higher temperatures is shown in Figure 2. Figure 2a shows the mass loss of PCDS calcined at different temperatures for 30 min. When the temperature is not higher than 1275 °C, the mass loss is less than 2.5%. When the temperature is 1300 °C, the mass loss is approximately 7.5%. When the temperature reaches 1325–1350 °C, the mass loss is about 12.5%, and when the temperature is not lower than 1375 °C, the mass loss is nearly 20%. Figure 2b shows the mass loss of PCDS calcined for different times at 1300 °C. When PCDS is calcined for 15 min, the mass loss is about 2.5%. When PCDS is calcined for 30–60 min, the mass loss is basically maintained at 7.5%. When PCDS is calcined for 75 min, the mass loss is close to 15%. In this study, the calcination time was fixed at 30 min, and additional PCDS was added to the proportion scheme according to the data in Figure 2a at each calcination temperature. (3) In the production of SAC, the three moduli (alkalinity coefficient: C_m, Al:S ratio: P and Al:Si ratio: N) [22] is usually used to control the clinker composition and adjust materials proportioning. The cement clinker involved in this study is HBSAC clinker, the C_2S content is usually higher than $C_4A_3\overline{S}$, and most of the clinkers contain residual $CaSO_4$, so C_m was designed to be slightly larger than 1.0, P not more than 3.82, and N not controlled.

Table 2. Design of the mineral composition of the clinkers and the proportion of the raw materials, wt.%.

Series	Cement	Preliminary ClinkerMinerals				Raw Materials			
		C_4AF	$C_4A_3\overline{S}$	C_2S	$CaSO_4$	PCDS	FA	CS	BX
GF-HBSAC-X	W$\overline{S}$-GF-HBSAC-DG	5	42	53	0	14.7	16.9	49.2	19.2
	GF-HBSAC	5	35	45	15	37.5	14.3	32.0	16.2
GF-HBSAC-Y	GF-HBSAC-15%								
	GF-HBSAC-10%	5	35	50	10	29.0	16.0	38.6	16.4
	GF-HBSAC-20%	5	35	40	20	47.2	9.9	23.2	19.7

Series	Cement	Final Clinker Minerals				
		C_4AF	$C_4A_3\overline{S}$	C_2S	$CaSO_4$	CaO
GF-HBSAC-X	W$\overline{S}$-GF-HBSAC-DG	5.27	42.15	52.47	0	0.11
	GF-HBSAC	5.09	34.24	45.14	15.45	0.08
GF-HBSAC-Y	GF-HBSAC-15%					
	GF-HBSAC-10%	5.23	36.12	48.52	10.08	0.05
	GF-HBSAC-20%	4.63	36.25	38.38	20.68	0.06

Note: The amount of PCDS in the table does not include any additional parts due to decomposition.

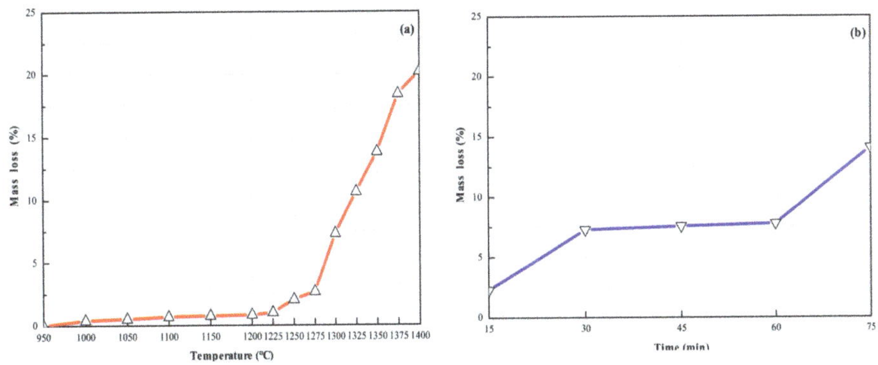

Figure 2. Mass loss of PCDS (**a**) at different calcination temperatures; (**b**) for different calcination time.

2.3. Preparation Process

The preparation of HBSAC clinker based on solid wastes was mainly divided into three steps: Grinding and molding, preheating and calcining, and cooling and regrinding, as shown in Figure 3.

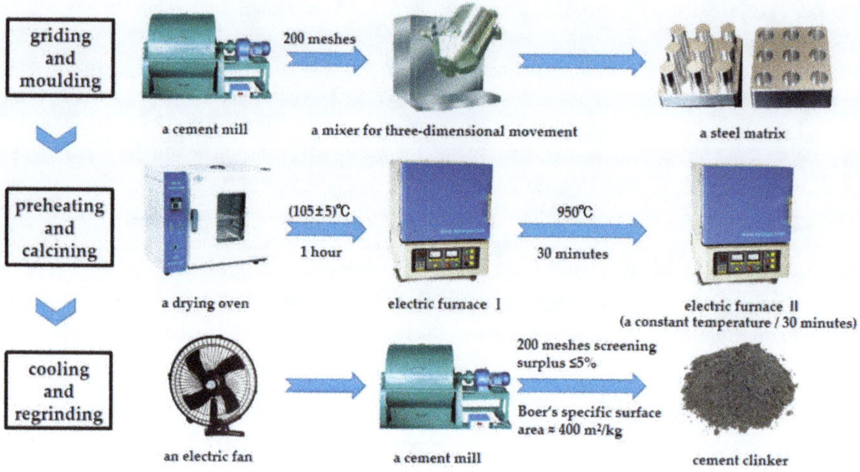

Figure 3. Preparation process of HBSAC clinker based on solid wastes.

Grinding and molding: All of the raw materials were ground by a cement mill to pass through a 200 mesh square hole sieve, and then mixed evenly by a mixer for three-dimensional movement, and finally pressed into a steel matrix to form cylindrical specimens of $\varphi \times h = 15 \times (15-20)$ mm.

Preheating and calcining: The samples were dried for 1 h in a drying oven at 105 ± 5 °C, and then preheated for 30 min in a high temperature electric furnace at 950 °C, and again calcined for 30 min in another high temperature electric furnace at a constant temperature.

Cooling and regrinding: The samples were taken out from the high temperature electric furnace, and then cooled by an electric fan, and finally ground to pass through a 200 mesh square hole sieve with a surplus of less than 5% or a Boer's specific surface area of about 400 m^2/kg.

2.4. Test Methods

-XRF: An X-ray fluorescence spectrometry (1800 type, Shimadzu Co., Kyoto, Japan) was used to analyze the chemical composition of the raw materials and clinkers. The test results were given in terms of oxide content.

-XRD: An X-ray diffraction instrument (D8 advance type, Bruker Co., Karlsruhe, Germany) was applied to detect mineral composition of the clinkers and hydration products. Its working conditions were as follows: Cu target, voltage at 40 kV, current at 40 mA, 2-Theta scanning ranges from 5° to 60°, step width of 0.02°, and a residence time of 0.05 s. In order to quantify phases of the clinkers, QXRD analysis was conducted by using EXPGUI-GSAS [23] software to refine XRD patterns of the clinkers. It should be noted that all samples were in powder form, and the samples of hydration products were obtained from 20 × 20 × 20 mm cement pastes prepared under the condition of water:cement ratio of 0.50.

-SEM: A scanning electron microscopy (JSM-7500F type, JEOL Co., Ltd., Tokyo, Japan) was adopted to characterize the micromorphology of the clinker minerals and hydration products with a working voltage of 2.0 kV. The samples of the clinkers were in powder, and the samples of the hydration products were in pieces, which were all sprayed with gold before observed. The samples of hydration products were also obtained from 20 × 20 × 20 mm cement pastes prepared under the condition of water:cement ratio of 0.50.

-Isothermal calorimetric: An eight-channel isothermal conduction calorimeter (TAM Air type, Thermometric AB Co., Jafalla, Sweden) was used to measure the hydration heat of cements. In the test, the temperature was set at 25 °C, the water:cement ratio was fixed to 0.5, and the hydration heat were recorded over a period of 72 h.

-Thermal analysis: A comprehensive thermal analyzer (SDT Q600 type, TA Instruments Co., New Castle, DE, USA.) was used to conduct differential thermal analysis of hydration products. The temperature varied from room temperature (approximately 20 °C) to 800 °C at a heating rate of 20 °C/min with a flux of N_2. The samples of the hydration products were in powder, and the samples of hydration products were the same as the samples used for XRD analysis.

-Particle size distribution: A laser particle size analyzer (Rise-2006 type, Jinan Rise Science and Technology Co., Ltd., Jinan, China) was adopted to test the particle size distributions of cements or clinkers and alcohol was selected as dispersing medium.

-Physical and mechanical properties test: The water requirement of normal consistency and setting time were determined according to the "Test methods for water requirement of normal consistency, setting time and soundness of the Portland cement" (GB/T 1346-2011, China). The mechanical strength was tested according to the "Method of testing cements-Determination of strength" (GB/T 17671-1999, China) and the "sulfoaluminate cement" (GB 20472-2006, China). The size of cement mortars were 40 × 40 × 160 mm, and the water requirement was determined by controlling the fluidity to 170 mm. The size of cement pastes were 20 × 20 × 20 mm, which were prepared under the condition of water:cement ratio of 0.35 and 0.50. After preparation, all samples were placed in a curing room with a temperature of 20 ± 1 °C and a relative humidity of no less than 90% for 6 h, then demolded and cured in water until different ages.

3. Results and Discussion

3.1. Preparation of HBSAC Based on Solid Wastes

3.1.1. Series GF-HBSAC-X Cement

Figure 4 shows the XRD patterns of GF-HBSAC and W$\overline{S}$-GF-HBSAC-DG clinkers at different calcination temperatures. It can be seen from the figure that there are obvious $CaSO_4$ diffraction peaks in the XRD patterns of GF-HBSAC clinker, while there was no $CaSO_4$ diffraction peaks in the XRD patterns of W$\overline{S}$-GF-HBSAC-DG clinker, which is consistent with the designed mineral composition of two clinkers. In addition, both clinkers contain minerals such as $C_4A_3\overline{S}$, β-C_2S and $C_4A_{2.85}F_{0.15}\overline{S}$, among which $C_4A_3\overline{S}$ and β-C_2S are the main mineral components of SAC, and $C_4A_{2.85}F_{0.15}\overline{S}$ is a solid solution formed by the solid solution of iron element in $C_4A_3\overline{S}$. According to the relationship between the intensity of the diffraction peaks, the optimal calcination temperature of the two clinkers is 1300 °C.

For each of two clinkers prepared at 1300 °C, the chemical composition was tested by XRF, and the three moduli was calculated. The results are all listed in Table 3. It is worth noting that the equations for calculating the three moduli should be adjusted because of the existence of $C_4A_{2.85}F_{0.15}\overline{S}$ and $CaSO_4$ in clinker, and the adjusted equations are as follows (3)–(5).

$$C_m = \{\omega(CaO) - 0.7\omega(TiO_2) - 1.70[\omega(SO_3) - 0.26\omega(Al_2O_3)]\}/[0.73\omega(Al_2O_3) + 1.87\omega(SiO_2)]. \quad (3)$$

$$P = \omega(Al_2O_3)/\omega(SO_3). \quad (4)$$

$$N = \omega(Al_2O_3)/\omega(SiO_2). \quad (5)$$

The mineral content of two clinkers was analyzed by QXRD [19,24,25] using GSAS software. The results are listed in Table 4. Comparing Tables 2 and 4, the actual mineral content of each clinker is close to the designed mineral content, and the content error is within a reasonable range. Figure 5 shows the refined XRD pattern of GF-HBSAC clinker. As can be seen from this figure, the fitting peak shape (surrounded by black circle) is similar to the experimental peak shape (surrounded by red solid line), and R_{wp} = 8.95 < 10 and χ^2 = 1.978, so it can be considered that the refined result is reliable. For this reason, the reliability of materials proportioning and calcination system has also been verified.

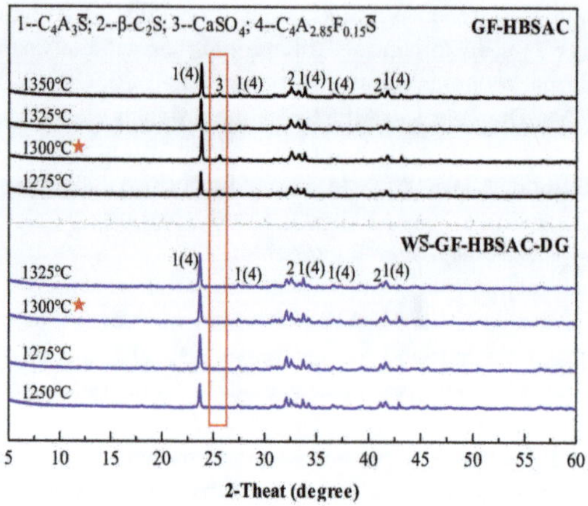

Figure 4. X-ray diffraction (XRD) patterns of the Series GF-HBSAC-X cement clinkers at different temperatures.

Table 3. Chemical composition and the three moduli of the Series GF-HBSAC-X cement clinkers.

Clinker	Chemical Composition (%)								The Three Moduli		
	CaO	Al_2O_3	SiO_2	Fe_2O_3	SO_3	MgO	TiO_2	Σ	C_m	P	N
GF-HBSAC	46.75	18.24	14.13	2.47	12.93	2.30	0.60	97.42	1.02	1.41	1.29
$W\overline{S}$-GF-HBSAC-DG	48.06	21.91	16.77	2.52	5.82	2.52	0.71	97.84	1.00	3.76	1.31

Table 4. Mineral composition of the Series GF-HBSAC-X cement clinkers.

Clinker	Mineral Composition (%)			
	$C_4A_3\overline{S}$	β-C_2S	$CaSO_4$	Others
GF-HBSAC	36.27	50.96	12.62	0.15
$W\overline{S}$-GF-HBSAC-DG	43.18	56.53	0.00	0.29

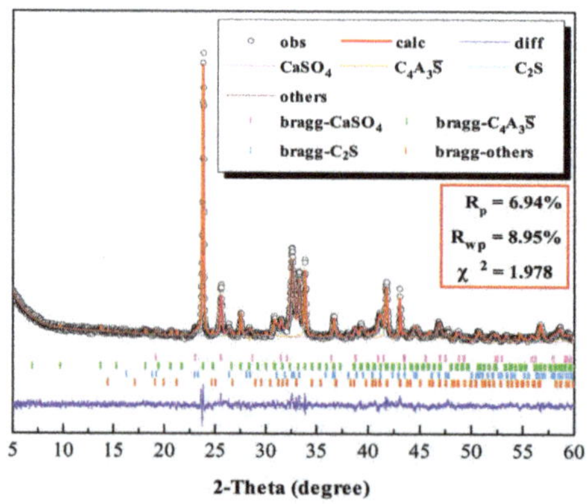

Figure 5. Refined X-ray diffraction (XRD) pattern of GF-HBSAC clinker by using GSAS.

In combination with the results of QXRD analysis, $W\overline{S}$-GF-HBSAC-DG cement was prepared by mixing 84% $W\overline{S}$-GF-HBSAC-DG clinker with 16% DG on the principle of equal $CaSO_4$ and $C_4A_3\overline{S}$ contents in GF-HBSAC and $W\overline{S}$-GF-HBSAC-DG cement. Figure 6 shows the XRD patterns of the two cements. It can be seen from the figure that there are obvious diffraction peaks of β-C_2S and $C_4A_3\overline{S}$ ($C_4A_{2.85}F_{0.15}\overline{S}$) in both cements, but the diffraction peaks of $CaSO_4$ component show that the $CaSO_4$ component in GF-HBSAC cement is anhydrous $CaSO_4$, while the $CaSO_4$ component in W$\overline{S}$-GF-HBSAC-DG cement is DG. Figure 7 shows the SEM micromorphology of the two cements. In the two cements, blocky and granular β-C_2S and polygonal and tabular $C_4A_3\overline{S}$ are all observed, and the obvious difference lies in the morphology of CaSO4 component. The $CaSO_4$ component in GF-HBSAC cement is rectangular while the $CaSO_4$ component is lamellar in W$\overline{S}$-GF-HBSAC-DG cement, which is consistent with the crystal structure of anhydrous $CaSO_4$ and DG [25–29].

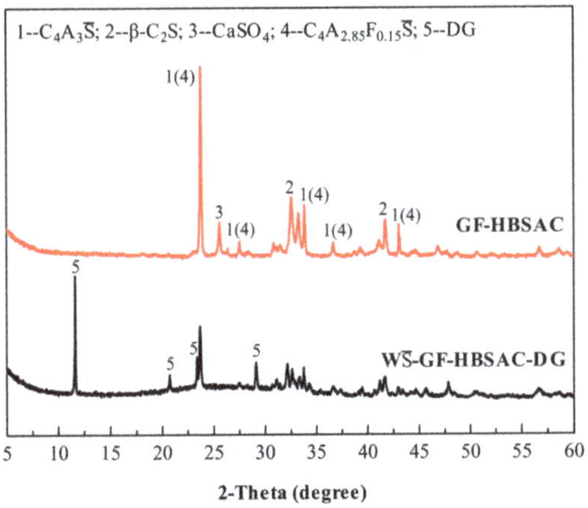

Figure 6. X-ray diffraction (XRD) patterns of the Series GF-HBSAC-X cement.

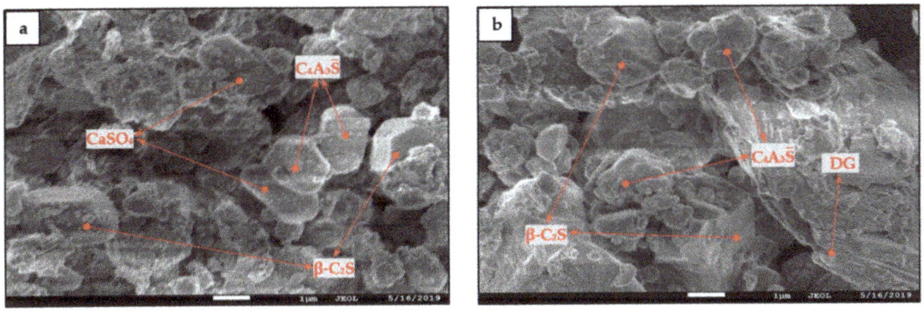

Figure 7. SEM micromorphology of the Series GF-HBSAC-X cement: (**a**) GF-HBSAC cement; (**b**) W$\overline{S}$-GF-HBSAC-DG cement.

3.1.2. Series GF-HBSAC-Y Cement

Figure 8 shows the XRD patterns of the Series GF-HBSAC-Y cement clinkers prepared at different calcination temperatures. It is; thus, clear that the formation temperature of cement clinkers with coexisting minerals, including $C_4A_3\overline{S}$, β-C_2S, and $CaSO_4$, is affected by residual $CaSO_4$ content in clinker. For GF-HBSAC-10% clinker, GF-HBSAC-15% clinker, and GF-HBSAC-20% clinker, the formation temperature ranges from 1225 to 1350 °C, 1250 to 1350 °C, and 1275 to 1350 °C, respectively. That is to say, with the increase of residual $CaSO_4$ content in clinker, the lower limit temperature of clinker formation gradually increases. The reason for this phenomenon lies in the formation and decomposition of $C_5S_2\overline{S}$ [15,19,30], and the highest temperature of $C_5S_2\overline{S}$ in the above three clinkers reaches 1200, 1225, and 1250 °C, respectively. For the optimal calcination temperature and the upper limit temperature of clinker formation, there is little difference among the three clinkers. According to the relationship between the intensity of the diffraction peaks, the optimal calcination temperature of the three clinkers is 1300 °C. It is reasonable to set the upper limit temperature at 1350 °C, at which the decomposition of various minerals has been very large, especially the $CaSO_4$.

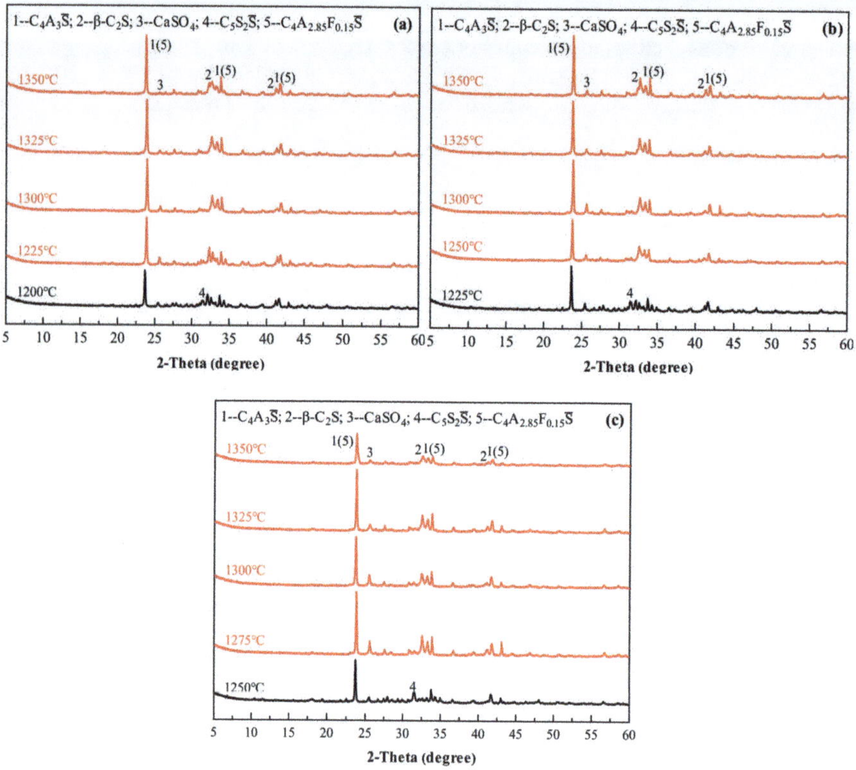

Figure 8. X-ray diffraction (XRD) patterns of the Series GF-HBSAC-Y cement clinkers at different temperatures: (**a**) GF-HBSAC-10%; (**b**) GF-HBSAC-15%; (**c**) GF-HBSAC-20%.

For each of the three clinkers prepared at 1300 °C, the chemical composition was tested by XRF, and the three moduli was calculated according to Equations (3)–(5), and the results are listed in Table 5. The mineral content of the three clinkers was still analyzed by QXRD using GSAS software and the results are listed in Table 6.

Table 5. Chemical composition and the three moduli of the Series GF-HBSAC-Y cement clinkers.

Clinker	Chemical Composition (%)								The Three Moduli		
	CaO	Al_2O_3	SiO_2	Fe_2O_3	SO_3	MgO	TiO_2	Σ	C_m	P	N
GF-HBSAC-10%	47.23	18.09	15.46	2.44	9.86	2.42	0.64	96.14	1.02	1.83	1.17
GF-HBSAC-15%	46.75	18.24	14.13	2.47	12.93	2.30	0.60	97.42	1.02	1.41	1.29
GF-HBSAC-20%	45.83	18.25	12.63	2.15	15.88	3.00	0.77	98.51	1.01	1.15	1.44

Figure 9 shows SEM micromorphology of the Series GF-HBSAC-Y cement clinkers. In general, the mineral micromorphology of the three clinkers seems to be the same, but there are some differences in detail. As shown in Figure 9a, the boundary between crystals is obvious and the crystal size of clinker minerals is large. In addition, in Figure 9b, the boundary between crystals is still obvious but the crystal size of clinker minerals becomes smaller; and in Figure 9c, the boundary between crystals is no longer clear, which is consistent with the results of reference [31].

Table 6. Mineral composition of the Series GF-HBSAC-Y cement clinkers.

Clinker	Mineral Composition (%)			
	$C_4A_3\overline{S}$	β-C_2S	$CaSO_4$	Others
GF-HBSAC-10%	36.13	55.28	8.34	0.25
GF-HBSAC-15%	36.27	50.96	12.62	0.15
GF-HBSAC-20%	35.91	46.36	17.65	0.08

Figure 9. SEM micromorphology of the Series GF-HBSAC-Y cement: (**a**) GF-HBSAC-10%; (**b**) GF-HBSAC-15%; (**c**) GF-HBSAC-20%.

3.2. *Effect of $CaSO_4$ Type on Properties of HBSAC Based on Solid Wastes*

3.2.1. Effect of $CaSO_4$ Type on Water Requirement of Normal Consistency and Setting Time

As shown in Figure 10, the water requirement of normal consistency of GF-HBSAC cement reaches 37%, whereas that of W$\overline{S}$-GF-HBSAC-DG cement is only 29%. The main reason for the obvious difference is that the residual $CaSO_4$ in GF-HBSAC clinker dissolves, diffuses, and crystallizes into DG after encountering water, resulting in an increase in water demand. In terms of setting time, the setting time of GF-HBSAC cement is longer than that of W$\overline{S}$-GF-HBSAC-DG cement, which is attributed to the different hydration activity of residual $CaSO_4$ in clinker and DG. Studies in [32,33] have shown that DG and hemihydrate gypsum can more effectively promote the dissolution of $C_4A_3\overline{S}$ and the formation of AFt than anhydrous $CaSO_4$, and the reason is that slow dissolution of anhydrous $CaSO_4$ leads to insufficient supply of calcium and sulfate ions.

3.2.2. Effect of $CaSO_4$ Type on Hydration Heat

Figure 11 shows the hydration heat curves of the Series GF-HBSAC-X cement. As can be seen from Figure 11a, there are three exothermic peaks in the heat flow curve of GF-HBSAC cement and W$\overline{S}$-GF-HBSAC-DG cement, which is the same as the findings in [19]. The first exothermic peak is generated by the dissolution heat of cement, the second exothermic peak by the hydration of $C_4A_3\overline{S}$, and the third exothermic peak by the secondary hydration of $C_4A_3\overline{S}$ [34]. However, the time and peak value of exothermic peaks are different between the two cements, especially for the second and third exothermic peaks. The time of the second and third exothermic peaks of W$\overline{S}$-GF-HBSAC-DG cement are 0.93 and 2.84 h, respectively, which are earlier than that of GF-HBSAC cement, which are 1.95 and 7.97 h, respectively. The peak values of the second and third exothermic peaks of W$\overline{S}$-GF-HBSAC-DG cement are 0.015 and 0.006 W/g, respectively, which are larger than that of GF-HBSAC cement, which are 0.012 and 0.005 W/g, respectively. Considering that the two cements have similar particle size distribution, as shown in Figure 12, the effect of cement fineness on the hydration rate can be negligible, so the exothermic process mainly depends on the cement composition. According to the results of QXRD analysis of clinkers and the method of cement preparation in Section 3.1.1, the difference between the two cements lies in the form of $CaSO_4$ composition and the content of β-C_2S, while the hydration of β-C_2S is very slow. Therefore, the difference in Figure 11a is supposed to be caused by the difference between residual $CaSO_4$ and DG [32,33]. The above results show that the residual $CaSO_4$ in clinker can prolong the duration of induction period and reduce the hydration exothermic rate in the early stage, especially within 8 h. As shown in Figure 11b, the cumulative heat released from GF-HBSAC cement is gradually higher than that of W$\overline{\text{S}}$-GF-HBSAC-DG cement after 9.33 h. At 72 h, the cumulative heat released from GF-HBSAC cement reaches 197 J/g, while that of W$\overline{\text{S}}$-GF-HBSAC-DG cement is only 161 J/g, which is due to the combination of residual $CaSO_4$ in GF-HBSAC clinker and water to release heat.

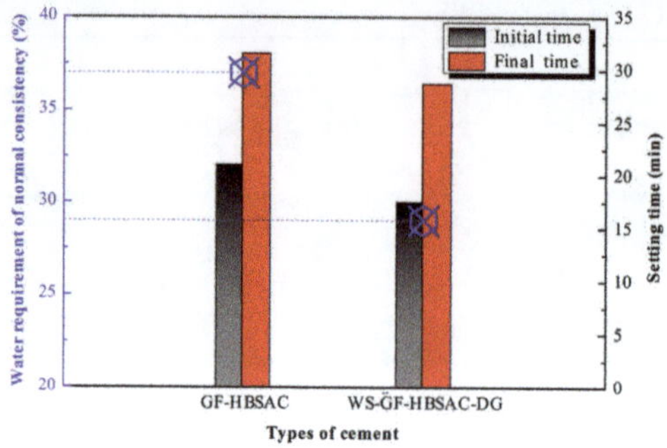

Figure 10. Water requirement of normal consistency and setting time of the Series GF-HBSAC-X cement.

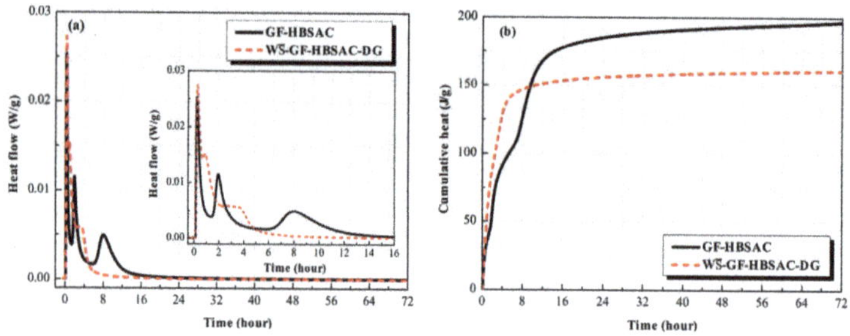

Figure 11. Hydration heat curves of the Series GF-HBSAC-X cement: (**a**) Heat flow curve; (**b**) cumulative heat curve.

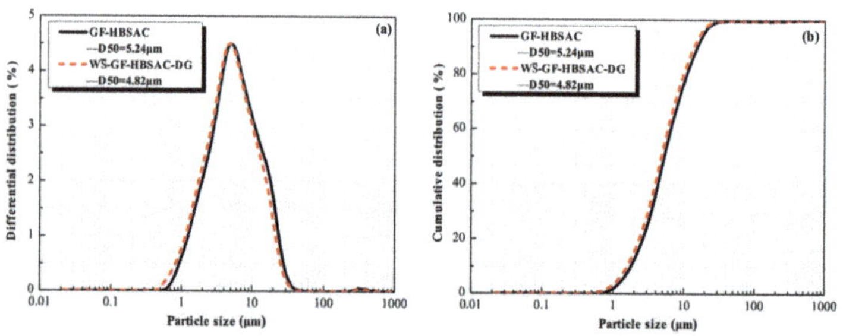

Figure 12. Particle size distribution curves of the Series GF-HBSAC-X cement: (**a**) Differential distribution; (**b**) cumulative distribution.

3.2.3. Effect of $CaSO_4$ Type on Mechanical Strength

It can be seen from Figure 13 that the compressive strength of GF-HBSAC and W$\overline{\text{S}}$-GF-HBSAC-DG cement pastes under different water: cement ratios is comparatively consistent. There is little difference in the compressive strength between the two cement pastes during 1 to 3 days. Since

7 days, the compressive strength of GF-HBSAC cement pastes continuously increased, while that of $W\overline{S}$-GF-HBSAC-DG cement pastes showed a decrease in strength, which may be caused by the expansion and cracking of the cement pastes, as shown in Figure 14, and similar results have been found in [35], indicating that DG is likely to exceed the optimal content of gypsum. From this point of view, residual $CaSO_4$ can completely replace DG to participate in hydration reaction. Compared with DG, residual $CaSO_4$ can still ensure the stable development of cement strength when the design content of residual $CaSO_4$ in clinker is 15%, reflecting the advantage of residual $CaSO_4$ in HBSAC clinker based on solid wastes, at the same time, residual $CaSO_4$ in clinker is also beneficial to the utilization of PCDS.

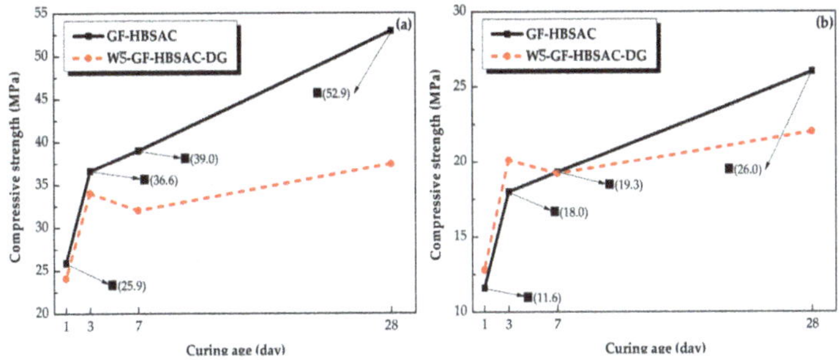

Figure 13. Compressive strength of the Series GF-HBSAC-X cement pastes under different water:cement ratios: (**a**) Water:cement ratio of 0.35; (**b**) water:cement ratio of 0.50.

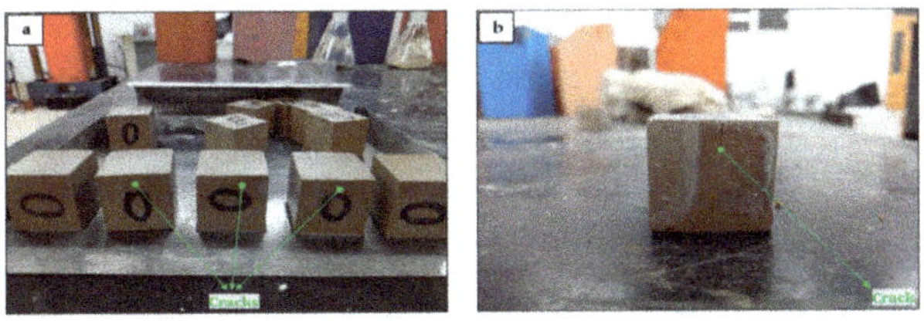

Figure 14. Development of cracks in $W\overline{S}$-GF-HBSAC-DG cement pastes with curing ages: (**a**) Seven days; (**b**) 28 days.

3.2.4. Effect of $CaSO_4$ Type on Hydration Products

As can be seen from Figure 15, the development of hydration products of GF-HBSAC and $W\overline{S}$-GF-HBSAC-DG cement during 1 to 28 days is basically similar. In terms of the types of hydration products, the hydration products of both cements are mainly composed of AFt, indicating that residual $CaSO_4$ can promote the hydration of $C_4A_3\overline{S}$ to form AFt as DG. From the formation process of hydration products, AFt formed in a large amount in both cements hydration at 1 day, while $C_4A_3\overline{S}$ and $CaSO_4$ were consumed in large quantities. The diffraction peaks of $C_4A_3\overline{S}$ tended to be stable at 3 days and still existed at 28 days, and the diffraction peaks of β-C_2S did not change significantly during the whole process, indicating that the hydration of $C_4A_3\overline{S}$ begins rapidly in the early stage, while the hydration of β-C_2S is relatively slow, which is consistent with the viewpoint of [35,36].

As can be seen from Figure 16, the position of the dehydration peaks in DTG curves of GF-HBSAC and $W\overline{S}$-GF-HBSAC-DG cement is the same, indicating that the hydration products of the two cements are the same. Generally speaking, the dehydrated phase at 100–150 °C is AFt, the dehydrated phase at 130–160 °C is AFm or C_2ASH_8 ($2CaO{\cdot}Al_2O_3{\cdot}SiO_2{\cdot}8H_2O$), and the dehydrated phase at 240–270 °C is AH_3. The reason why AFm or C_2ASH_8 and AH_3 are not reflected in XRD patterns is probably due to the low content and poor crystallization.

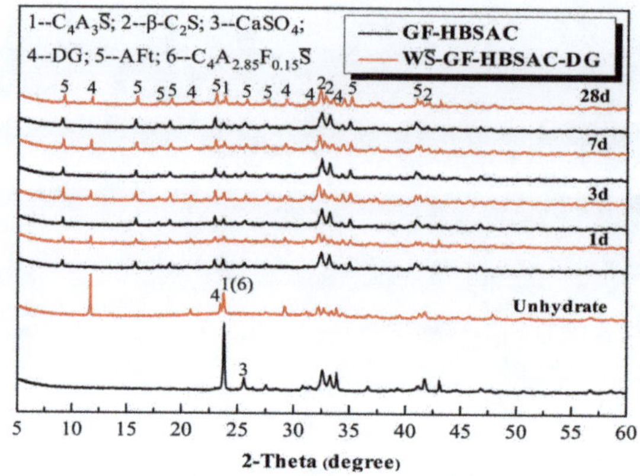

Figure 15. X-ray diffraction (XRD) patterns of the Series GF-HBSAC-X cement hydration products at different curing ages.

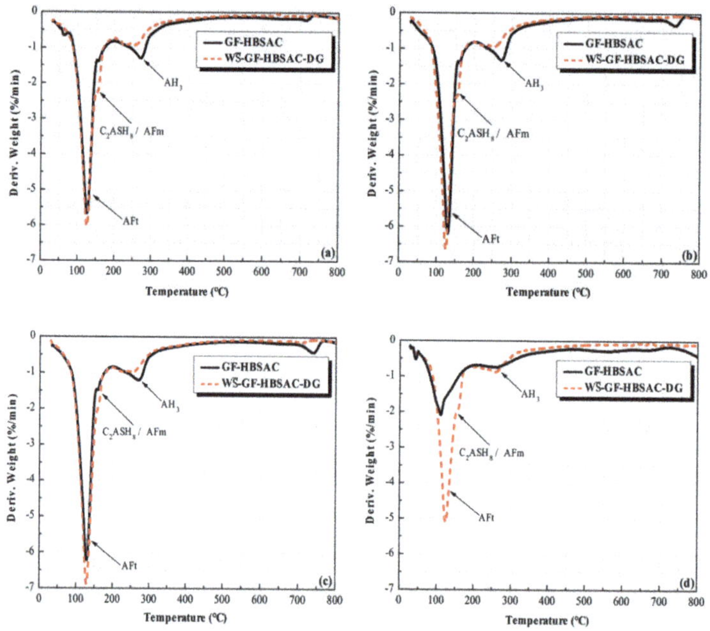

Figure 16. DTG curves of the Series GF-HBSAC-X cement hydration products at different curing ages: (**a**) One day; (**b**) 3 days; (**c**) 7 days; (**d**) 28 days.

From the main hydration product, the AFt content of W$\bar{S}$-GF-HBSAC-DG cement hydration products is higher than that of GF-HBSAC cement during 1 to 28 days, which can be used to explain the development of cement pastes strength, as shown in Figures 13b and 14. From other hydration products, the AH_3 content of GF-HBSAC cement hydration products is higher than that of W$\bar{S}$-GF-HBSAC-DG cement, and the AFm or C_2ASH_8 content of GF-HBSAC cement hydration products is lower than that of W$\bar{S}$-GF-HBSAC-DG cement. According to the hydration reaction of Equations (1) and (2), the AH_3 content is supposed to increase with the increase of AFt or AFm content, but now the decrease of AH_3 content with the increase of AFt content means that AH_3 may be further hydrated and consumed. The hydration reaction of Equation (6) has been mentioned in [35–37]. Now, the increase of AFm or C_2ASH_8 content with the decrease of AH_3 content further confirms the possibility of this hydration reaction. Therefore, compared with the residual $CaSO_4$ in clinker, the additional DG is more likely to promote the hydration of C_2S.

$$C_2S + AH_3 + 5H \rightarrow C_2ASH_8. \quad (6)$$

Figures 17 and 18 show the SEM micromorphology of the two cements hydration products at different curing ages. According to the comparison, the two cements hydration products are basically the same, which are composed of needle-shaped AFt and cotton-shaped gel (including AH_3, C-S-H, etc.). The gel is wrapped around AFt and plays the role of caulking and gluing, and the structure of AFt combined with gel provides the necessary strength for cement pastes. With the hydration reaction proceeding, AFt gradually develops from fine needle to thick rod. During 1 to 28 days, the AFt of W$\bar{S}$-GF-HBSAC-DG cement hydration products seems to be stronger than that of GF-HBSAC cement hydration products, and fine needle-shaped AFt appears in the W$\bar{S}$-GF-HBSAC-DG cement hydration products at 7 days, which is initially determined as the newly formed AFt at the expansion cracking site. Therefore, it can be seen that the development of hydration products, especially the formation of AFt, coincides with the development of strength.

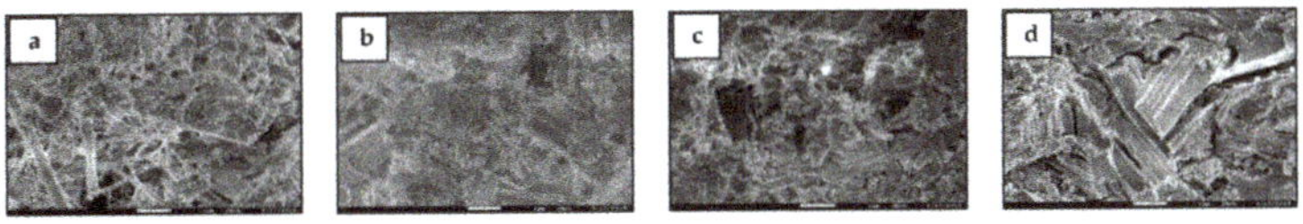

Figure 17. SEM micromorphology of GF-HBSAC cement hydration products at different curing ages: (**a**) One day; (**b**) 3 days; (**c**) 7 days; (**d**) 28 days.

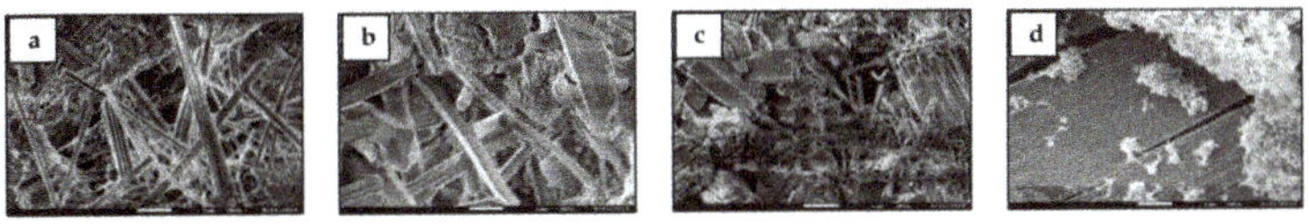

Figure 18. SEM micromorphology of W$\bar{S}$-GF-HBSAC-DG cement hydration products at different curing ages: (**a**) One day; (**b**) 3 days; (**c**) 7 days; (**d**) 28 days.

3.3. Effect of $CaSO_4$ Content on Properties of HBSAC Based on Solid Wastes

3.3.1. Effect of $CaSO_4$ Content on Water Requirement of Normal Consistency and Setting Time

From Figure 19 it can be seen that, with the increase of residual $CaSO_4$ content in clinker, the water requirement of normal consistency tends to decrease, and the setting time tends to shorten. It is well known that the effect of $CaSO_4$ content on the water requirement of normal consistency and setting time is related to the hydration of $C_4A_3\bar{S}$ and the formation of AFt. With the increase of $CaSO_4$ content, the hydration speed of $C_4A_3\bar{S}$ is accelerated [35,38], and the formation of AFt is also accelerated, which leads to the reduction of water demand and shortening of setting time. However, excessive $CaSO_4$ content may cause the formation of AFt too fast and too much so that AFt wraps on the surface of clinker mineral particles and affects the hydration reaction to continue, which results in the reverse change of increasing water demand and prolonging setting time. In this study, the water requirement of normal consistency and setting time do not change in reverse regularity when the $CaSO_4$ content increases, indicating that the increase of $CaSO_4$ content within the design content range of 10%–20% is always conducive to the hydration of $C_4A_3\bar{S}$.

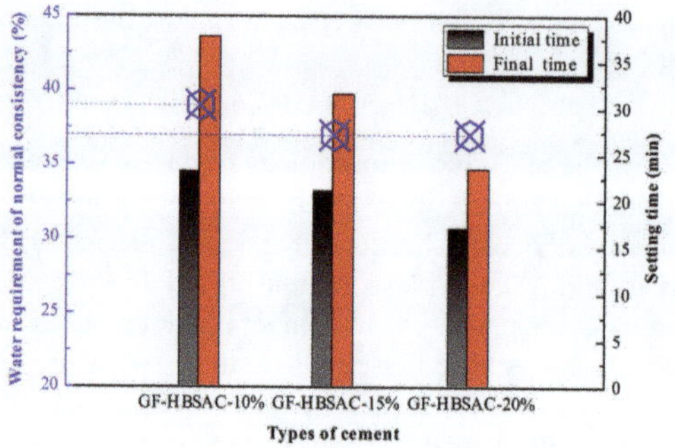

Figure 19. Water requirement of normal consistency and setting time of the Series GF-HBSAC-Y cement.

3.3.2. Effect of $CaSO_4$ Content on Hydration Heat

Figure 20 shows the hydration heat curves of the Series GF-HBSAC-Y cement. As can be seen from Figure 20a, the characteristics of heat flow curves of the three cements are similar, but the time and peak value of exothermic peaks are different. The first exothermic peaks of all three cements occur in about 10 min, and the peak value increases with the increase of $CaSO_4$ content in clinker. For the second exothermic peak, the time appears earlier with the increase of $CaSO_4$ content, and the duration and peak value increases with the increase of $CaSO_4$ content. As for the third exothermic peak, the difference between the three cements is more obvious. With the increase of $CaSO_4$ content, the time of exothermic peak appears earlier substantially, and the peak value increases first and then decreases, and the duration decreases gradually. It is concluded that the increase of $CaSO_4$ content promotes the hydration of $C_4A_3\overline{S}$. Figure 18b shows the cumulative heat curve of GF-HBSAC cement with different $CaSO_4$ contents. In the range of 0–26 h, the cumulative heat of three cements is as follows: GF-HBSAC-20% > GF-HBSAC-15% > GF-HBSAC-10%. In the range of 26–32 h, the cumulative heat of the three cements is as follows: GF-HBSAC-20% > GF-HBSAC-10% > GF-HBSAC-15%. After 32 h, the cumulative heat of three cements is as follows: GF-HBSAC-10% > GF-HBSAC-20% > GF-HBSAC-15%. It can be seen that the increase of residual $CaSO_4$ content is beneficial to the hydration of $C_4A_3\overline{S}$ in the early stage of hydration, but it also brings larger hydration heat. With the hydration reaction proceeding, the hydration degree of GF-HBSAC-10% cement which reacts slowly in the early stage increases, and the cumulative heat increases to the highest at 3 days, while that of GF-HBSAC-15% and GF-HBSAC-20% cement are basically the same.

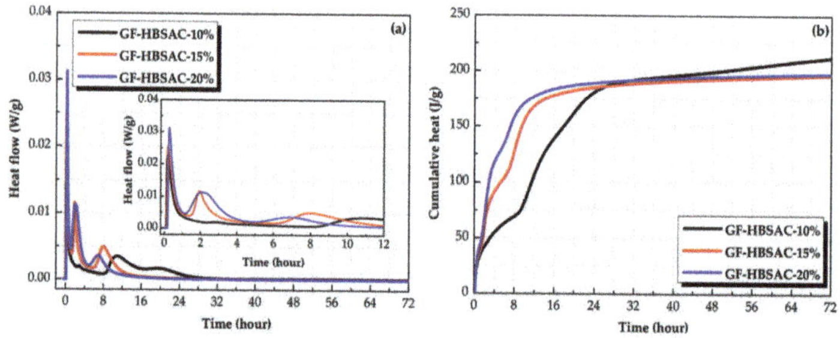

Figure 20. Hydration heat curves of the Series GF-HBSAC-Y cement. (**a**) Heat flow curve; (**b**) cumulative heat curve.

Based on the above analysis, GF-HBSAC-15% cement can not only maintain a good hydration rate, but also release less heat, which is the most prominent among the three cements with different $CaSO_4$ contents.

3.3.3. Effect of $CaSO_4$ Content on Mechanical Strength

Figure 21a shows the development of bending strength of the Series GF-HBSAC-Y cement. From this figure, it can be seen that the bending strength increases with the increase of $CaSO_4$ content during 1 to 7 days, but it no longer follows the same development rule during 7 to 28 days, and the relationship between the bending strength of three cements is GF-HBSAC-20% > GF-HBSAC-10% > GF-HBSAC-15%. The reason is that the strength growth is mainly due to the hydration of $C_4A_3\overline{S}$ to form AFt within 7 days, and the increase of $CaSO_4$ content can accelerate the hydration rate of $C_4A_3\overline{S}$. During 7 to 28 days, for GF-HBSAC-20% and GF-HBSAC-15% cement, the hydration of $C_4A_3\overline{S}$ basically ended. In addition, the strength growth is dominated by the hydration of β-C_2S in this stage. Therefore, the bending strength of GF-HBSAC-10% cement increases fastest.

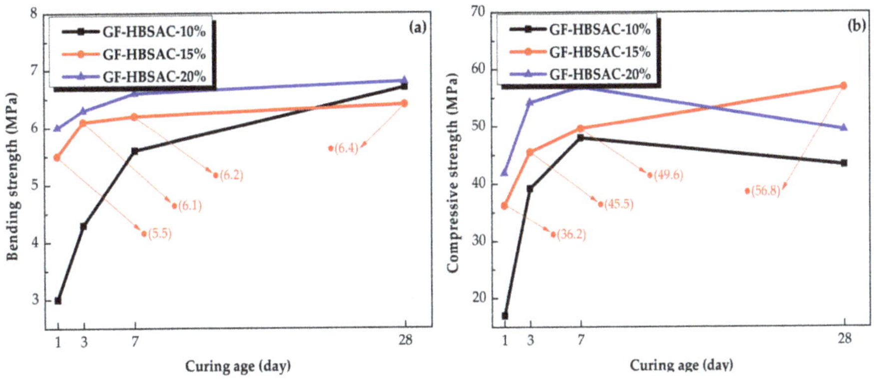

Figure 21. Mechanical strength of the Series GF-HBSAC-Y cement at different curing ages: (**a**) Bending strength; (**b**) compressive strength.

Figure 21b shows the development of compressive strength of the Series GF-HBSAC-Y cement. From this figure, it can be seen that the development rule of compressive strength with $CaSO_4$ content is the same as that of bending strength during 1 to 7 days, but there is a new development rule during 7 to 28 days. The compressive strength of both GF-HBSAC-10% and GF-HBSAC-20% cement shows a tendency of retrogression, and the relationship between the compressive strength of three cements is GF-HBSAC-15% > GF-HBSAC-20% > GF-HBSAC-10%. The reason is that the strength growth is mainly due to the hydration of $C_4A_3\overline{S}$ to form AFt within 7 days, and the increase of $CaSO_4$ content can accelerate the hydration rate of $C_4A_3\overline{S}$. During 7 days to 28 days, the decrease in strength of GF-HBSAC-10% cement may be attributed to the formation of AFm [5,16,39], while the decrease in strength of GF-HBSAC-20% cement may be due to the micro-cracks, which are caused by the pressure of AFt crystals formed rapidly in the early stage continuous growth, and the pressure of gel filling cement paste pore in the later stage [35,40,41].

Based on the above analysis, the mechanical strength of GF-HBSAC-15% cement develops steadily, which is the most outstanding among the three cements.

3.3.4. Effect of $CaSO_4$ Content on Hydration Products

It can be seen from Figure 22 that the main hydration product of GF-HBSAC-Y cement is AFt. At 1 day, with the increase of $CaSO_4$ content, the diffraction peaks of $C_4A_3\overline{S}$ weaken and the diffraction peaks of AFt enhance, indicating that the increase of $CaSO_4$ content promotes the hydration of $C_4A_3\overline{S}$ to form AFt. AFm appears in GF-HBSAC-10% cement, which should be the reaction of $C_4A_3\overline{S}$ with water to form low sulfur ettringite. At 3 days, for all three cements, the diffraction peaks of $C_4A_3\overline{S}$ continue to weaken and the diffraction peaks of AFt continue to enhance. During 3 to 7 days, the diffraction peaks of GF-HBSAC-10% cement hydration products vary the most, indicating that the hydration degree of GF-HBSAC-10% cement begins to increase. At 7 days, the diffraction peaks of GF-HBSAC-15% and GF-HBSAC-20% cement hydration products have not changed much, but for the GF-HBSAC-10% cement, the diffraction peaks of AFt begin to decrease, and the diffraction peaks of AFm reappear, which is considered to be transformed from AFt under the condition of low $CaSO_4$ content. At 28 days, the diffraction peaks of AFm continue to enhance in GF-HBSAC-10% cement, which may lead to the weakening of cement pastes and the retrogression of strength. In addition, the diffraction peaks of β-C_2S decrease considerably, and C_2ASH_8 formed in GF-HBSAC-10% cement, which should be the hydration product of reaction between β-C_2S and AH_3 according to Equation (6), but not observed in hydration products of the other two cements. Considering that the β-C_2S content in GF-HBSAC-10% cement is the highest, this phenomenon may be related to the β-C_2S content in cement.

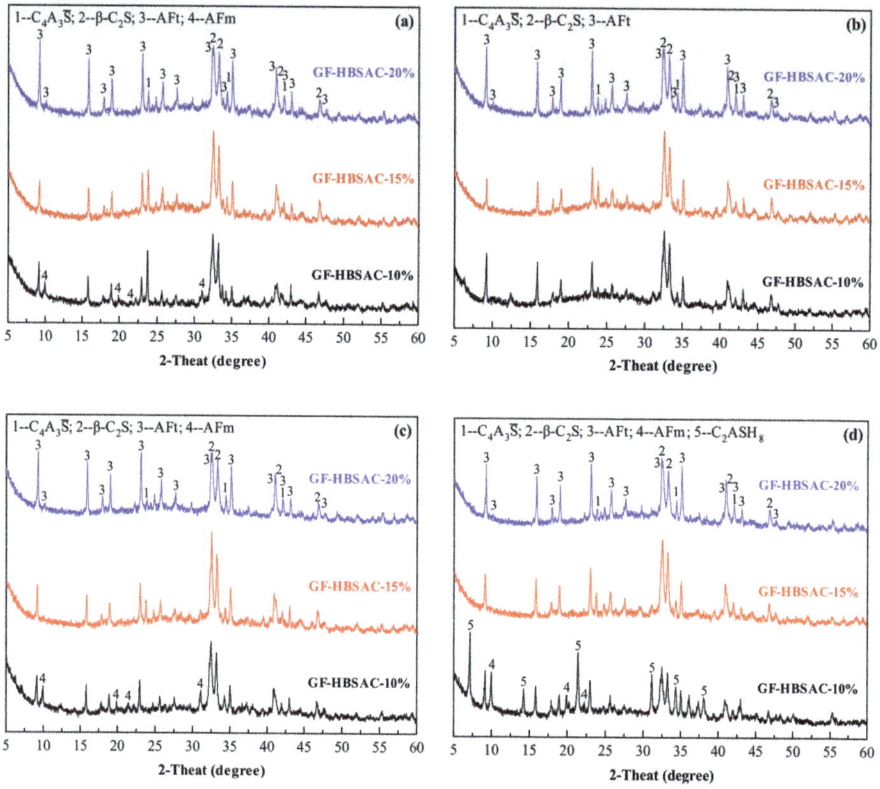

Figure 22. X-ray diffraction (XRD) patterns of the Series GF-HBSAC-Y cement hydration products at different curing ages: (**a**) One day; (**b**) 3 days; (**c**) 7 days; (**d**) 28 days.

As can be seen from Figure 23, the position of the dehydration peaks in DTG curves of three cements is the same, indicating that the hydration products of three cements are the same. The hydration products are mainly composed of AFt, AFm, or C_2ASH_8 and AH_3. For AFt, the content increases with the increase of $CaSO_4$ content during 1 to 28 days, which is coincided with the change rule reflected by XRD patterns and further indicates that the increase of $CaSO_4$ content promotes the hydration of $C_4A_3\overline{S}$. For AH_3, the change rule with $CaSO_4$ content varies in different curing ages. At 1 day, the AH_3 content increases first and then decreased with the increase of $CaSO_4$ content. Considering that hydration reaction at this stage is dominated by hydration of $C_4A_3\overline{S}$, the increase of $CaSO_4$ content should promote hydration of $C_4A_3\overline{S}$ to form more AH_3, but the AH_3 content of GF-HBSAC-20% cement decreases instead, which should be caused by the consumption of part of AH_3. Combining Equation (6), it can be considered that β-C_2S in cement of high $CaSO_4$ content takes the lead in hydration. During 3 to 7 days, the hydration degree of GF-HBSAC-10% cement increases, while the hydration of the other two cements slows down, so the AH_3 content in GF-HBSAC-10% cement reached the highest level. In this stage, the reason why the diffraction peaks of hydration products such as C_2ASH_8 do not appear in XRD patterns may be related to the low hydration degree of β-C_2S and the low crystallization degree of C_2ASH_8. At 28 days, the AH_3 content of the three cements decreases, and the AH_3 content of GF-HBSAC-10% cement decreases to the lowest, which may be due to the greatest hydration degree of β-C_2S and the best crystallization degree of C_2ASH_8 in GF-HBSAC-10% cement, corresponding to the presence of C_2ASH_8 detected in XRD analysis. For AFm or C_2ASH_8, since the dehydration peaks of the two hydration products overlap, it is difficult to make a clear change rule of each product with $CaSO_4$ content through DTG curves, which needs further study.

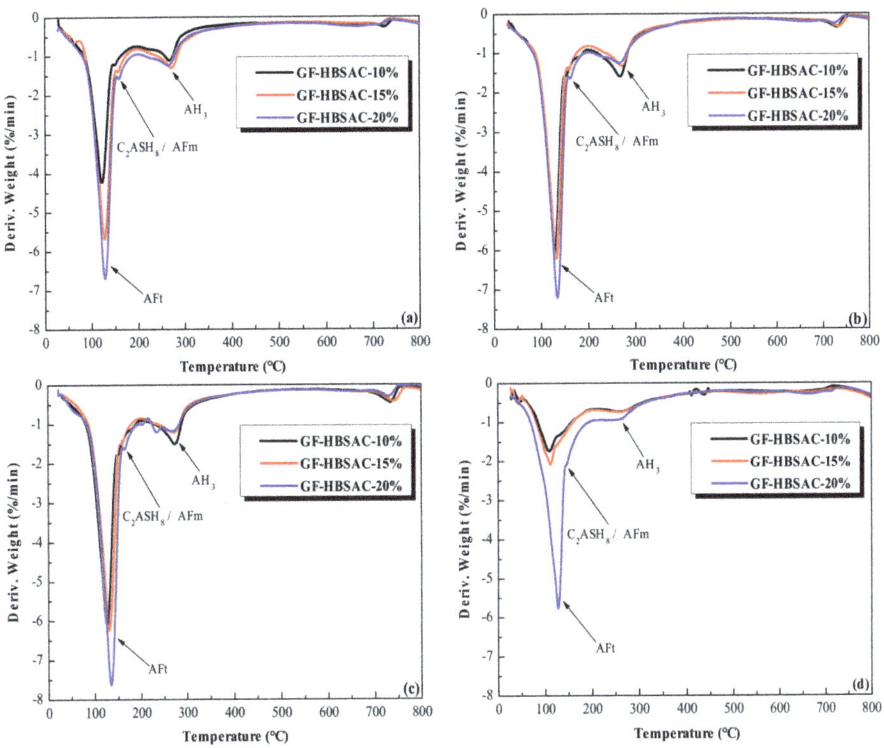

Figure 23. DTG curves of the Series GF-HBSAC-Y cement hydration products at different curing ages: (**a**) One day; (**b**) 3 days; (**c**) 7 days; (**d**) 28 days.

Figures 24 and 25 show SEM micromorphology of GF-HBSAC-10% cement and GF-HBSAC-20% cement hydration products at different curing ages, respectively. Combined with Figure 17, for the three cements, the types of hydration products are basically similar, but the development of hydration products varies with different $CaSO_4$ contents. For example, the AFt in GF-HBSAC-20% cement has developed into a thick rod and the structure of cement paste is relatively dense at 1 day, while the AFt in the other two cements is thin needles and the structure of cement paste is relatively sparse. Besides, lamellar AFm is found in the hydration products of GF-HBSAC-10% cement at 1 and 7 days, which is consistent with the results of XRD analysis. In the hydration products of GF-HBSAC-20% cement at 28 days, cracks around AFt can be clearly found, which should be caused by the expansion pressure because of the development of AFt.

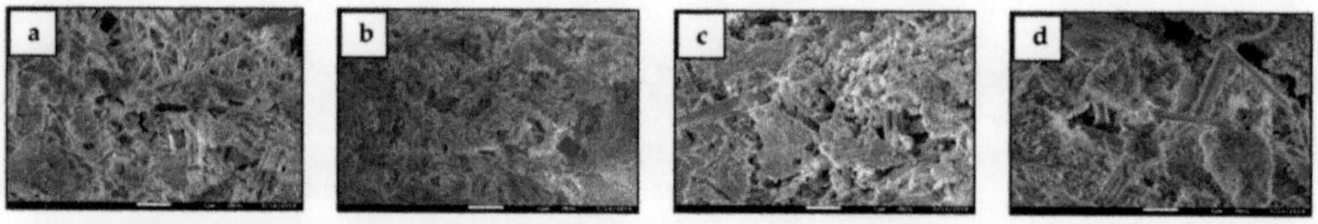

Figure 24. SEM micromorphology of GF-HBSAC-10% cement hydration products at different curing ages: (**a**) One day; (**b**) 3 days; (**c**) 7 days; (**d**) 28 days.

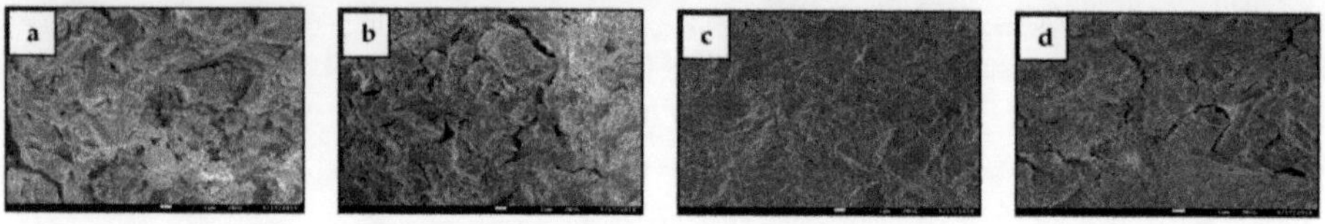

Figure 25. SEM micromorphology of GF-HBSAC-20% cement hydration products at different curing ages: (**a**) One day; (**b**) 3 days; (**c**) 7 days; (**d**) 28 days.

4. Conclusions

It is feasible to prepare HBSAC cement containing residual $CaSO_4$ or no residual $CaSO_4$ by using solid wastes including PCDS, FA, CS, and BX. The utilization rate of solid wastes in raw materials can reach more than 80%. The application provides a new channel for the utilization of various industrial solid wastes, especially for the utilization of PCDS.

Residual $CaSO_4$ in clinker can take part in hydration instead of additional DG. Compared with additional DG, the residual $CaSO_4$ in clinker has both advantages and disadvantages on the properties of HBSAC cement based on solid wastes. In the hydration process, the residual $CaSO_4$ does not change the hydration products of HBSAC cement, and shows some advantages in strength. However, it increases the water requirement of normal consistency and hydration heat of cement, and prolongs the setting time. In addition, the promotion of residual $CaSO_4$ on the hydration of β-C_2S is not as strong as that of DG.

The residual $CaSO_4$ content affects the calcination of HBSAC cement clinkers based on solid wastes. With the increase of $CaSO_4$ content, the lower limit temperature of clinker formation increases gradually, but the optimal calcination temperature is not affected which is basically maintained at 1300 °C. The error between the actual and designed mineral content of clinkers prepared at 1300 °C is within a reasonable range. The micromorphology of clinkers also varies with the $CaSO_4$ content. Meanwhile, the crystal size of clinker becomes finer and the boundary between crystals becomes more blurred.

The residual $CaSO_4$ content also affects the hydration properties of HBSAC cement clinkers based on solid wastes. With the increase of $CaSO_4$ content, the water requirement of normal consistency decreases, the setting time shortens, and the hydration heat releases in advance. In addition, the bending strength and compressive strength enhance with the increase of $CaSO_4$ content during 1 to 7 days, but the bending strength of GF-HBSAC-10% cement still develops well during 7 to 28 days while that of the other two cements was basically unchanged, and the compressive strength of GF-HBSAC-15% cement shows a continuous development while that of the other two cements shows a retrogression. As for cement hydration products, the increase of $CaSO_4$ content does not change the types of major hydration products, and AFt develops better with the increase of $CaSO_4$ content. Meanwhile, AFm, as the secondary hydration product, is easy to produce in the process of cement hydration with lower $CaSO_4$ content. Besides, the increase of $CaSO_4$ content seems to promote the hydration of β-C_2S, which will consume part of AH_3 to form C_2ASH_8, but C_2ASH_8 has the better crystallization in cement with high β-C_2S content and low $CaSO_4$ content. Considering all aspects of analysis, the optimal design residue of $CaSO_4$ is 15%.

Author Contributions: Investigation, methodology, formal analysis, data curation, and writing—original draft, D.S.; project administration, conceptualization, and supervision, Q.L.; writing—review, editing and software, Y.G.; resources, visualization, and funding acquisition, G.Y.; validation, L.W. All authors have read and agreed to the published version of the manuscript.

Funding: Projects No. 51878366, 51578297, 51978353 and 51808310 from the National Natural Science Foundation of China, Projects No. ZR2019PEE007 from the Natural Science Foundation of Shandong Province.

Acknowledgments: The authors gratefully acknowledge the financial support of Projects No. 51878366, 51578297, 51978353 and 51808310 from the National Natural Science Foundation of China, Projects No. ZR2019PEE007 from the Natural Science Foundation of Shandong Province.

Conflicts of Interest: The authors declare no conflicts of interest.

Abbreviations

The following abbreviations are used in this manuscript:

HBSAC	high belite sulfoaluminate cement
SAC	sulfoaluminate cement
$C\bar{S}$	$CaSO_4$
$C\bar{S}H_2$	$CaSO_4{\cdot}2H_2O$
$C_4A_3\bar{S}$	$3CaO{\cdot}3Al_2O_3{\cdot}CaSO_4$
β-C_2S	$2CaO{\cdot}SiO_2$
$C_5S_2\bar{S}$	$4CaO{\cdot}2SiO_2{\cdot}CaSO_4$
C_4AF	$4CaO{\cdot}Al_2O_3{\cdot}Fe_2O_3$
f-CaO	free CaO
$C_4A_{2.85}F_{0.15}\bar{S}$	$3CaO{\cdot}2.85Al_2O_3{\cdot}0.15Fe_2O_3{\cdot}CaSO_4$
C_2ASH_8	$2CaO{\cdot}Al_2O_3{\cdot}SiO_2{\cdot}8H_2O$
AH_3	$Al_2O_3{\cdot}3H_2O$
AFt	$3CaO{\cdot}Al_2O_3{\cdot}3CaSO_4{\cdot}32H_2O$
AFm	$3CaO{\cdot}Al_2O_3{\cdot}CaSO_4{\cdot}12H_2O$
PCDS	petroleum coke desulfurization slag
FA	fly ash
CS	carbide slag
BX	bauxite
DG	dihydrate gypsum

References

1. Glasser, F.P.; Zhang, L. High-performance cement matrices based on calcium sulfoaluminate-belite compositions. *Cem. Concr. Compos.* **2001**, *31*, 1881–1886. [CrossRef]
2. Li, J.; Zhou, C.Y.; Yang, Y.J. Optimization of mineral composition of high belite-sulfoaluminate clinker. *J. Chin. Ceram. Soc.* **2012**, *40*, 1618–2012. (In Chinese)
3. El-Alfi, E.A.; Gado, R.A. Preparation of calcium sulfoaluminate-belite cement from marble sludge waste. *Constr. Build. Mater.* **2016**, *113*, 764–772. [CrossRef]
4. Xue, P.; Xu, A.J.; He, D.F.; Yang, Q.X.; Liu, G.Q.; Engström, F.; Björkman, B. Research on the sintering process and characteristics of belite sulfoaluminate cement produced by BOF slag. *Constr. Build. Mater.* **2016**, *122*, 567–576. [CrossRef]
5. Adolfsson, D.; Menad, N.; Viggh, E.; Björkman, B. Hydraulic properties of sulfoaluminate belite cement based on steelmaking slags. *Adv. Cem. Res.* **2007**, *19*, 133–138. [CrossRef]
6. Li, F.W. Research of Preparing Sulphate Aluminium Cement Using Calcium and Aluminum Residue and Low Grade Bauxite. Master's Thesis, Nanchang University, Nanchang, China, 2012. (In Chinese).
7. Wu, K.; Shi, H.S.; Guo, X.L. Utilization of municipal solid waste incineration fly ash for sulfoaluminate cement clinker production. *Waste Manag.* **2011**, *31*, 2001–2008. [CrossRef] [PubMed]
8. Shen, Y. Synthesis of Sulfate-Rich Belite Sulfoaluminate Cement with Phosphogypsum. Ph.D. Thesis, Chongqing University, Chongqing, China, 2013.

9. Huang, Y.B.; Qian, J.S.; Kang, X.J.; Yu, J.C.; Fan, Y.R.; Dang, Y.D. Belite-calcium sulfoaluminate cement prepared with phosphogypsum: Influence of P_2O_5 and F on the clinker formation and cement performances. *Constr. Build. Mater.* **2019**, *203*, 432–442. [CrossRef]
10. Wang, W.L.; Wang, X.J.; Zhu, J.P.; Wang, P.; Ma, C.Y. Experimental investigation and modeling of sulfoaluminate cement preparation using desulfurization gypsum and red mud. *Ind. Eng. Chem. Res.* **2013**, *52*, 1261–1266. [CrossRef]
11. Zhang, Z.Q. Fast-Setting and Quick-Hardening Belite sulfoaluminate Cement Clinker, Its Application and Production Process. Chinese Patent No. 201410416928.8, 25 August 2015. (In Chinese).
12. Sahu, S.; Majling, J. Preparation of sulfoaluminate belite cement from fly ash. *Cem. Concr. Res.* **1994**, *24*, 1065–1072. [CrossRef]
13. Li, H.X.; Agrawal, D.K.; Cheng, J.P.; Silsbee, M.R. Microwave sintering of sulfoaluminate cement with utility wastes. *Cem. Concr. Res.* **2001**, *31*, 1257–1261. [CrossRef]
14. Chen, I.A.; Juenger, M.C. Incorporation of coal combustion residuals into calcium sulfoaluminate-belite cement clinkers. *Cem. Concr. Compos.* **2012**, *34*, 893–902. [CrossRef]
15. Bullerjahn, F.; Schmitt, D.; Haha, M.B. Effect of raw mix design and of clinkering process on the formation and mineralogical composition of (ternesite) belite calcium sulfoaluminate ferrite clinker. *Cem. Concr. Res.* **2014**, *59*, 87–95. [CrossRef]
16. Rungchet, A.; Poon, C.; Chindaprasirt, P.; Pimraksa, K. Synthesis of low-temperature calcium sulfoaluminate-belite cements from industrial wastes and their hydration: Comparative studies between lignite fly ash and bottom ash. *Cem. Concr. Compos.* **2017**, *83*, 10–19. [CrossRef]
17. Xu, G.L. Research of Utilizing Coal Gangue in Shi Zuishan District to Produce Series of Sulfoaluminate Cements. Master's Thesis, Chengdu University of Technology, Chengdu, China, 2009. (In Chinese).
18. Singh, M.; Kapur, P.C. Preparation of calcium sulfoaluminate cement using fertiliser plant wastes. *J. Hazard. Mater.* **2008**, *157*, 106–113. [CrossRef] [PubMed]
19. Da Costa, E.B.; Rodríguez, E.D.; Bernal, S.A.; Provis, J.L.; Gobbo, L.A.; Kirchheim, A.P. Production and hydration of calcium sulfoaluminate-belite cements derived from aluminium anodising sludge. *Constr. Build. Mater.* **2016**, *122*, 373–383. [CrossRef]
20. Zhang, J.S.; Sui, Z.T.; Shen, Y.M.; Hui, Z.F.; An, H.Y.; Gao, F. Research of producing high silicon blite sulfur aluminate cement by titanium tailings. *Iron Steel Vanadium Titan.* **2004**, *25*, 41–47. (In Chinese)
21. Lv, S.Z.; Chen, X.M.; Lu, Z.Y.; Peng, Y.H. Study on high belite cement clinker calcination with ashes from circulating fluidized bed combustion. *New Build. Mater.* **2011**, *8*, 1–3. (In Chinese)
22. Wang, Y.M.; Su, M.Z.; Zhang, L. *Sulfoaluminate Cement*, 1st ed.; Press of Beijing University of Technology: Beijing, China, 1999. (In Chinese)
23. Toby, B.H. EXPGUI, a graphical user interface for GSAS. *J. Appl. Crystallogr.* **2001**, *34*, 210–213. [CrossRef]
24. Martín-Sedeño, M.C.; Cuberos, A.J.; Ángeles, G.; Álvarez-Pinazo, G.; Ordónez, L.M.; Gateshki, M.; Aranda, M.A. Aluminum-rich belite sulfoaluminate cements: Clinkering and early age hydration. *Cem. Concr. Compos.* **2010**, *40*, 359–369. [CrossRef]
25. Li, X.R.; Zhang, Y.; Shen, X.D.; Wang, Q.Q.; Pan, Z.G. Kinetics of calcium sulfoaluminate formation from tricalcium aluminate, calcium sulfate and calcium oxide. *Cem. Concr. Res.* **2014**, *55*, 79–87. [CrossRef]
26. Yang, S.Z.; Mo, S.B.; Zhao, S.L.; Song, H.T. A study on phase transition at high temperature and structure of anhydride. *Acta Mineral. Sin.* **1998**, *18*, 73–79. (In Chinese)
27. Chen, J.W.; Li, F.; Liu, Y.F.; Zhang, Z.L. Gypsum fibrosum and gypsum fibrosum preparatum by scanning electron microscopy. *Cent. South Pharm.* **2006**, *4*, 253–255. (In Chinese)
28. Yang, L.; Wang, W.; Liang, W.J.; Zhang, Y.S.; Zhang, F. Observation of gypsum fibrosum from different habitats by scanning electron microscopy. *Chin. Pharm.* **2015**, *18*, 326–328. (In Chinese)
29. Erbs, A.; Nagalli, A.; De Carvalho, K.Q.; Mymrin, V.; Passig, F.H.; Mazer, W. Properties of recycled gypsum from gypsum plasterboards and commercial gypsum throughout recycling cycles. *J. Clean. Prod.* **2018**, *183*, 1314–1322. [CrossRef]
30. Kacimi, L.; Cyr, M.; Clastres, P. Synthesis of α' L-C_2S cement from fly-ash using the hydrothermal method at low temperature and atmospheric pressure. *J. Hazard. Mater.* **2010**, *181*, 593–601. [CrossRef]
31. Lan, M.Z.; Jin, Y.L.; Ge, Z.X.; Liu, C.J.; Wang, J.F. Effect of free gypsum on firing of high belite calcium sulfoaluminate cement clinker. *Bull. Chin. Ceram. Soc.* **2018**, *37*, 3054–3059, 3065. (In Chinese)

32. Allevi, S.; Marchi, M.; Scotti, F.; Bertini, S.; Cosentino, C. Hydration of calcium sulfoaluminate clinker with additions of different calcium sulphate sources. *Mater. Struct.* **2016**, *49*, 453–466. [CrossRef]
33. García-Maté, M.; Angeles, G.; León-Reina, L.; Losilla, E.R.; Aranda, M.A.; Santacruz, I. Effect of calcium sulfate source on the hydration of calcium sulfoaluminate eco-cement. *Cem. Concr. Compos.* **2015**, *55*, 53–61. [CrossRef]
34. Jansen, D.; Spies, A.; Neubauer, J.; Ectors, D.; Goetz-Neunhoeffer, F. Studies on the early hydration of two modifications of ye'elimite with gypsum. *Cem. Concr. Res.* **2017**, *91*, 106–116. [CrossRef]
35. Morin, V.; Termkhajornkit, P.; Huet, B.; Pham, G. Impact of quantity of anhydrite, water to binder ratio, fineness on kinetics and phase assemblage of belite-ye'elimite-ferrite cement. *Cem. Concr. Res.* **2017**, *99*, 8–17. [CrossRef]
36. Jeong, Y.; Hargis, C.W.; Chun, S.; Moon, J. The effect of water and gypsum content on strätlingite formation in calcium sulfoaluminate-belite cement pastes. *Constr. Build. Mater.* **2018**, *166*, 712–722. [CrossRef]
37. Winnefeld, F.; Lothenbach, B. Hydration of calcium sulfoaluminate cements-experimental findings and thermodynamic modelling. *Cem. Concr. Res.* **2010**, *40*, 1239–1247. [CrossRef]
38. Winnefeld, F.; Barlag, S. Calorimetric and thermogravimetric study on the influence of calcium sulfate on the hydration of ye'elimite. *J. Therm. Anal. Calorim.* **2010**, *101*, 949–957. [CrossRef]
39. Sahu, S.; Havlica, J.; Tomkova, V.; Majling, J. Hydration behaviour of sulfoaluminate belite cement in the presence of various calcium sulphates. *Thermochim. Acta* **1991**, *175*, 45–52. [CrossRef]
40. Chen, I.A.; Hargis, C.W.; Juenger, M.C. Understanding expansion in calcium sulfoaluminate-belite cements. *Cem. Concr. Res.* **2012**, *42*, 51–60. [CrossRef]
41. Bizzozero, J.; Gosselin, C.; Scrivener, K.L. Expansion mechanisms in calcium aluminate and sulfoaluminate systems with calcium sulfate. *Cem. Concr. Res.* **2014**, *56*, 190–202. [CrossRef]

materials

MDPI

Article

Properties of Iron Bacteria Biofouling on Ni-P-rGO Coating under Flowing Conditions

Mingyang Sun, Zhiming Xu *, Zuodong Liu, Bingbing Wang and Huishuang Di

School of Energy and Power Engineering, Northeast Electric Power University, Jilin 132012, China; sunmingyang0313@163.com (M.S.); liuzuodong@neepu.edu.cn (Z.L.); wangbb@neepu.edu.cn (B.W.); dhs_0411@163.com (H.D.)

* Correspondence: xuzm@neepu.edu.cn

Received: 9 January 2020; Accepted: 6 February 2020; Published: 7 February 2020

Abstract: Biofouling on heat exchange devices can decrease heat transfer efficiency, corrode materials, and even lead safety accidents. Most heat exchange devices are made of carbon steel, which produces biofouling easily. In this paper, nickel-phosphorus-reduced graphene oxide (Ni-P-rGO) coating was prepared on carbon steel by electroless plating as a kind of advanced material to study the properties of iron bacteria biofouling under flowing conditions. The coating was analyzed via scanning electron microscopy and Raman spectroscopy. The properties of iron bacteria biofouling on carbon steel and Ni-P-rGO coating were then compared under flowing conditions. Compared with carbon steel, the asymptotic value of fouling resistance on the Ni-P-rGO coating significantly decreased. Additionally, the induction period and the time of reaching the asymptotic value greatly increased. The inhibition properties of biofouling of advanced materials Ni-P-rGO coating under different temperatures, flow velocities, and initial concentrations was also studied.

Keywords: biofouling; iron bacteria; nickel-phosphorus-reduced graphene oxide (Ni-P-rGO); induction period; fouling resistance

---

1. Introduction

Biofouling is a biofilm or an organic film that is formed by bacteria, algae, and other microorganisms, and their excreta deposited on a solid surface [1]. Biofouling may not only decrease the heat transfer efficiency but also corrode the materials and even lead to accidents [2]. Xu et al. [3] studied the characteristics of iron bacteria and slime-forming bacteria biofouling in plate heat exchangers. Trueba et al. [4] investigated the effect of biofilms on the heat transfer process in the seawater cooled stainless steel condenser surface. The results showed that when the biofilm was thinner, the heat transfer process was better. Chen et al. [5] explored the growth characteristics of microorganisms on the surface of the heat exchanger under the conditions of changing temperature, flow velocities, and microbial concentration.

In the past 20 years, modified surfaces have been widely used in various industries due to the properties of anti-fouling, anti-corrosion, and non-pollution. Additionally, there have been several reports on the modified surfaces to inhibit the biofouling. Shao et al. [6] found that the silver coating can inhibit the attachment of *Pseudomonas aeruginosa (PA01)*. Xu et al. [7] demonstrated that electroless Ni-Cu-P coating has excellent anti-biofouling properties. Liu and Zhao. [8] found that the coating of Ni-P-ploy tetra fluoro ethylene (PTFE) exhibits excellent antimicrobial properties and reduced the *Escherichia (XA90)* adhesion by 95% as compared to stainless steel 304.

Because carbon materials have good superior chemical stability and high surface area [9], several researchers have added carbon materials to the electroless plating solution of Ni-P coating, and the resultant coating has good wear resistance, corrosion resistance, and antifouling. Tran et al. [10] produce high-performance CNT (carbon nano tube) /Au/Cu composite wires via sputtering and electroplating.

At the CNT volume fraction of more than 20%, the composite wires were lightweight and possessed the combined properties of effective strength and high electrical conductivity. Wu et al. [11] showed that the Ni-P-graphene oxide (GO) coating on the mild steel surface possesses the best anti-wear and microhardness properties when 40 mg/L graphene oxide is added into the electroless plating solution. Qian et al [12] found that the G-Ni-P composite plating increased the Vickers hardness and Young's modulus as compared with the Ni-P coating. Matjie et al. [13] prepared a modified diamond-like carbon (DLC) coating that reduced the adhesion of aluminum silicate deposit by 97% as compared with uncoated stainless steel. Shao et al. [14] added graphene oxide to reinforce the Ni-P coating material in order to inhibit the adhesion of *Staphylococcus aureus (ATCC 6538)*. Lee et al. [15] prepared the Ni-P/diamond, Ni-P/graphene, and Ni-P/diamond/graphene coating, and the Ni-P/diamond/graphene coating displayed the best hardness, corrosion resistance, and wear resistance in 3.5 wt % NaCl solution.

Although there has been significant work on biofouling and modified surfaces, there have been limited reports about the properties of biofouling on the nickel-phosphorus-reduced graphene oxide (Ni-P-rGO) coatings. The experimental conditions of the modified surface to inhibit fouling was almost static, as were the experimental conditions under flow conditions. Herein, Ni-P-rGO coating was prepared by electroless plating on carbon steel, and the iron bacteria was used as the scaling bacteria to study the biofouling properties of iron bacteria on Ni-P-rGO coating under flow conditions.

2. Experimental

2.1. Experimental Setup

Figure 1 shows the schematic of the experimental setup, where the solid line is the flow channel and the dotted line is the connection line of the data acquisition system. The experimental setup consisted of a low-temperature medium circulation circuit, a high-temperature medium circulation circuit, an experimental section, and a data acquisition system. During the experiment, the high temperature work fluid and low temperature work fluid exchanged heat in the experimental section via circulating pumps and flow meters, and their inlet and outlet temperatures were recorded through the data acquisition system.

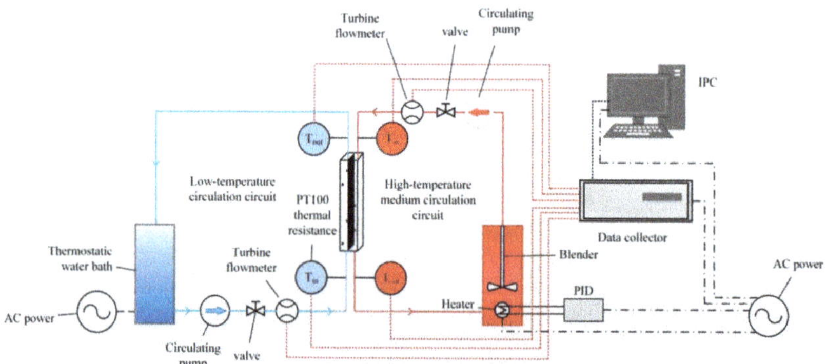

Figure 1. Schematic of experimental setup.

2.2. Experimental Principle and Method

The heat absorption by the low temperature working fluid during operation is equal to the heat release by the high temperature working fluid [3]:

$$\begin{gathered}\phi_1 = \phi_2 \\ q_{m1}c_{p1}\left(t_1'' - t_1'\right) = q_{m2}c_{p2}\left(t_2' - t_2''\right)\end{gathered} \tag{1}$$

where ϕ_1 is the heat release by the high temperature working fluid in W; ϕ_2 is the heat absorption by the low temperature working fluid in W; c_{p1} and c_{p2} are the specific heat of high temperature working fluid and low temperature working fluid, respectively, at constant pressure in kJ (kg K)$^{-1}$; q_{m1} and q_{m2} are the mass fluid velocity of the high temperature working fluid and low temperature working fluid in kg s^{-1}, respectively; and t_1', t_1'', t_2', and t_2'' are the inlet and outlet temperatures of the high temperature working fluid and low temperature working fluid in K.

Hence, the heat transfer coefficient can be calculated by:

$$k = \frac{\phi}{A\Delta t_m} = \frac{q_{m1}c_{p1}\left(t_1''-t_1'\right)+q_{m2}c_{p2}\left(t_2'-t_2''\right)}{2A\Delta t_m},\quad \Delta t_m = \frac{\left(t_2''-t_1'\right)-\left(t_2'-t_1''\right)}{ln\frac{t_2''-t_1'}{t_2'-t_1''}} \tag{2}$$

The fouling resistance R_f is defined as:

$$R_f = \frac{1}{k} - \frac{1}{k_0} \tag{3}$$

where A is the heat exchange surface in m^2; Δt_m is the logarithmic mean temperature difference in K; and k_0 and k are the overall heat transfer coefficient of clean (unfouled) surfaces and fouled surfaces in W (m^2 K)$^{-1}$.

2.3. Stability Verification of the Experimental Setup

In order to verify the stability of the experimental setup, repeatability experiments were conducted. The experiments were performed by taking distilled water as the circulating working fluid, high temperature working fluid at 65 °C, and 0.35 m/s and low temperature working fluid at 30 °C, increasing the flow velocity of the low temperature working fluid from 0.15 m/s to 0.35 m/s. The overall heat transfer coefficients for the repeatability experiments are shown in Figure 2. Under the conditions of changing the flow velocity, the maximum difference in the heat transfer coefficient of the two experiments was lower than 1%, which verifies the stability of the experimental setup.

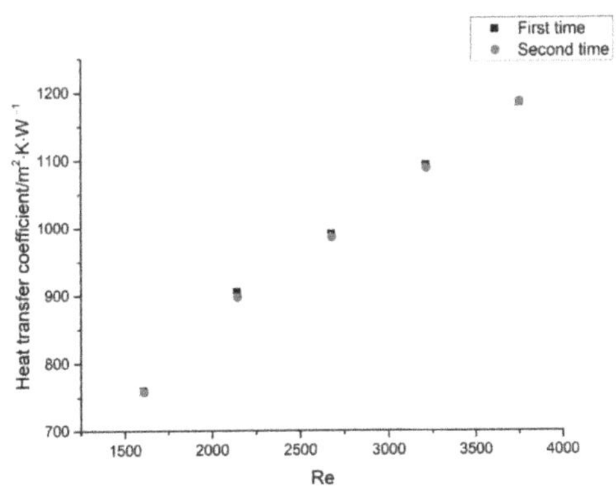

Figure 2. Heat transfer coefficients of the experiment setup at different flow velocities.

2.4. Coating Preparation and Iron Bacteria Culture

The Ni-P-rGO coating material was prepared on Q235 carbon steel (500 mm × 50 mm × 0.5 mm) via electroless plating. Before electroless plating, the substrates were first polished with an 800-grit waterproof abrasive paper and a 600-grit waterproof abrasive paper to make the surface smooth and flat, respectively. The oil was then removed using alkali washing and the rust was removed using

20% sulfuric acid for 1 min. Finally, the substrates were placed into 10% sulfuric acid for 1 min to activate the surface. After each step, the carbon steel surface was cleaned with distilled water. Table 1 lists the electroless bath formulation composition and operating conditions. The pH of the electroless bath solution was adjusted to 4.8 by adding ammonium hydroxide. The electroless plating process lasted 120 min. As shown in Figure 3, after successful plating, the coating was analyzed via scanning electron microscope and Raman spectroscopy. Figure 3a,b displays the morphology of carbon steel and Ni-P-rGO coating under a scanning electron microscope. The surface of the carbon steel was uneven and had several scratches and the Ni-P-rGO coating showed greater flatness than carbon steel in the microstructure. However, there were several inlaid nodules in the Ni-P-rGO coating, confirming the presence of graphene with a high specific surface area to stimulate the nucleation [13]. Figure 3c shows the Raman spectrum of the Ni-P-rGO coating. There were three evident peaks at 1350 cm^{-1}, 1580 cm^{-1}, and 2700 cm^{-1}, which correspond to the D band, G band, and 2D band. The D band is assigned to the breathing mode of the k-point phonons with A_{1g} symmetry [16] and it associates with the edge defects of graphene. The G band introduces the E_{2g} phonon of the carbon sp^2 atoms, which is due to the highly ordered graphite [17]. The 2D band was generated by double phonon double resonance, which is closely related to the band structure of graphene [18].

Table 1. The electroless bath solution composition and operating conditions.

Nickel Sulfate	25 g/L
Sodium Hypophosphite	30 g/L
Citric Acid	15 g/L
Lactic Acid	16 g/L
Sodium Acetate	11 g/L
Potassium Iodide	0.01 g/L
Graphene	40 mg/L
OP-10	Appropriate
pH	4.8
Temperature	83 °C

The iron bacteria used in this study was isolated and purified from the slime of the cooling tower in the power plant. The composition of liquid medium for iron bacteria is listed in Table 2. Before inoculating the iron bacteria, the pH of medium was adjusted to 6.8 and sterilized using autoclave for 30 min. After inoculation, the medium was placed into an incubator for 72 h at 30 °C.

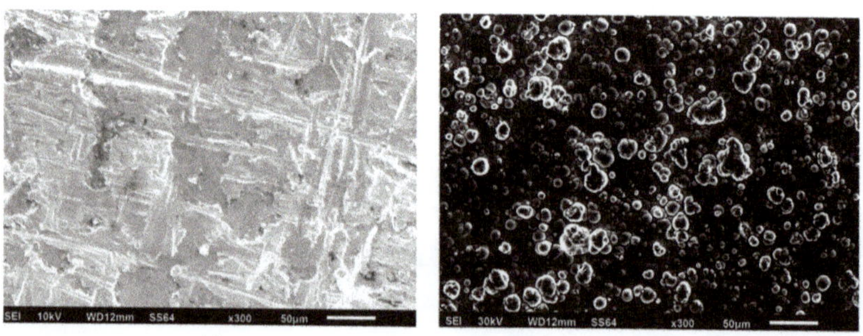

(a) (b)

Figure 3. *Cont.*

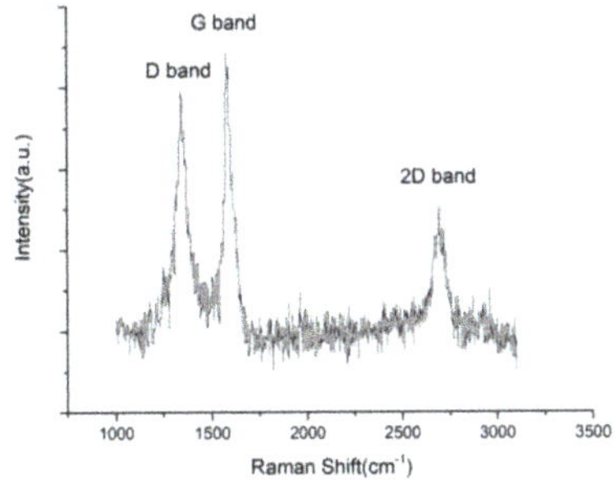

(c)

Figure 3. SEM morphology of carbon steel (**a**), nickel-phosphorus-reduced graphene oxide (Ni-P-rGO) coating (**b**), and Raman spectrum of rGO in Ni-P-rGO coatings (**c**).

Table 2. Medium composition for iron bacteria.

Ammonium Ferric Citrate	10 g/L
Sodium nitrate	0.5 g/L
Dipotassium phosphate	0.5 g/L
Calcium chloride	0.2 g/L
Magnesium sulfate	0.5 g/L
Ammonium sulfate	0.5 g/L

2.5. Experimental Progress

Before adding the iron bacteria into the thermostatic water bath, adjustment of the operation conditions was necessary. The fluid velocity was adjusted by the flow regulating valve and monitored via the flowmeter. The hot water was heated by the electric heater and controlled by the temperature controller. When the experimental conditions were stable, the iron bacteria was added into the thermostatic water bath. The data acquisition system collected data four times per minute. The collected data, such as inlet and outlet temperatures and flow velocity, were transmitted to the computer for storage. The data were processed after finishing the experiment. During the experiment, the concentration of iron bacteria in water was measured every 12 h using a spectrophotometer.

3. Results and Discussions

3.1. Comparison of Properties of Iron Bacteria Biofouling between Carbon Steel and Ni-P-rGO Coating

The experiments of carbon steel and Ni-P-rGO coating were performed when the conditions were with a low temperature working fluid at 30 °C, flow velocity at 0.25 m/s, and initial concentration of iron bacteria at 48.91×10^{10} CFU/mL, and with the high temperature working fluid at 65 °C, the flow velocity was at 0.35 m/s. The results are shown in Figure 4. As shown in the figure, the asymptotic value of the fouling resistance of Ni-P-rGO coating was much smaller than the carbon steel. Moreover, the fouling resistance of carbon steel increased rapidly as compared with the Ni-P-rGO coating. There was no induction period on carbon steel and it only took 90 h to reach the asymptotic value of fouling resistance. However, the Ni-P-rGO coating had an induction period of about 40 h, and the rate of fouling growth was extremely slow and took 175 h to reach the asymptotic value of fouling thermal resistance. This period of time was almost twice as long as the carbon steel. The changes in the concentration of iron bacteria during the experiment are shown in Figure 5. The concentration of iron bacteria in the carbon steel experiment increased significantly at the initial stage and decreased rapidly after reaching the peak at 36 h. However, at the initial stage of the Ni-P-rGO coating experiment, the concentration of iron bacteria almost remained invariable and decreased rapidly after 80 h. This phenomenon can be explained by the theory that the iron element in carbon steel provides for the growth and reproduction of iron bacteria. At the beginning of the experiment, the iron bacteria greatly multiplied and adhered to the surface of carbon steel and the carbon steel was corroded to produce a large number of corrosion products. When iron bacteria multiplied too much, the surface of carbon steel was gradually covered by iron bacteria biofouling and the iron bacteria could not contact the carbon steel to obtain nutrients. At the same time, large amounts of iron bacteria also consumed large amounts of nutrients. Therefore, bacteria decayed rapidly and eventually reached an asymptotic value of fouling resistance. In contrast, the iron bacteria on the coating could not obtain any nutrients and made contact with graphene in the coating during the corrosion process. Graphene has a special hexagonal structure that can puncture the cytomembrane of iron bacteria [19]. The propagating iron bacteria and the decaying iron bacteria reached a dynamic balance; hence, the concentration of iron bacteria was almost invariable at the initial stage of the experiment. After a period of corrosion by the iron bacteria, increasingly more graphene was exposed and there were less nutrients in the water; therefore, the iron bacteria decayed rapidly and reached an asymptotic value of fouling resistance. Figure 6 shows the macrograph and micrograph of the carbon steel and coating surface after the experiment. As the macrograph shows, the surface of carbon steel was almost covered with red-brown iron bacterial fouling. However, the surface of the Ni-P-rGO coating had only sporadic iron bacteria biofouling. As the micrograph shows, there was not only a thicker layer of biofouling, but also a significant iron bacteria colony on the carbon steel. However, on the coating of the Ni-P-rGO, biofouling was almost absent and the coating surface was almost free of corrosion damage.

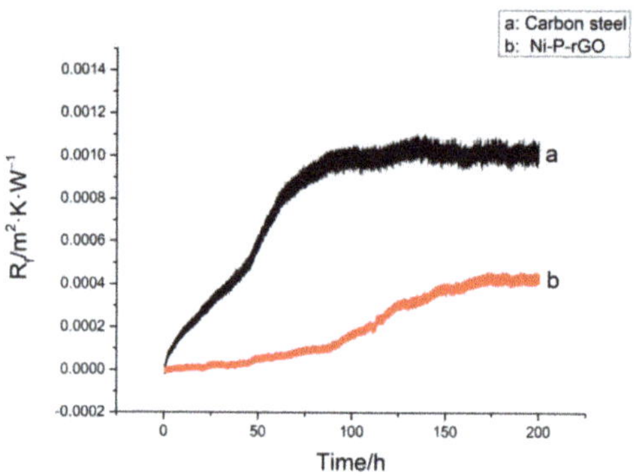

Figure 4. The properties of the iron bacteria biofouling between carbon steel and Ni-P-rGO coating.

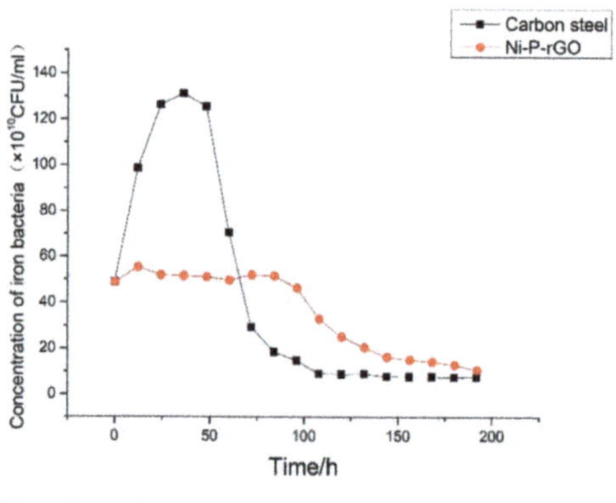

Figure 5. Variation of iron bacteria concentration during the experiment in the low temperature working fluid.

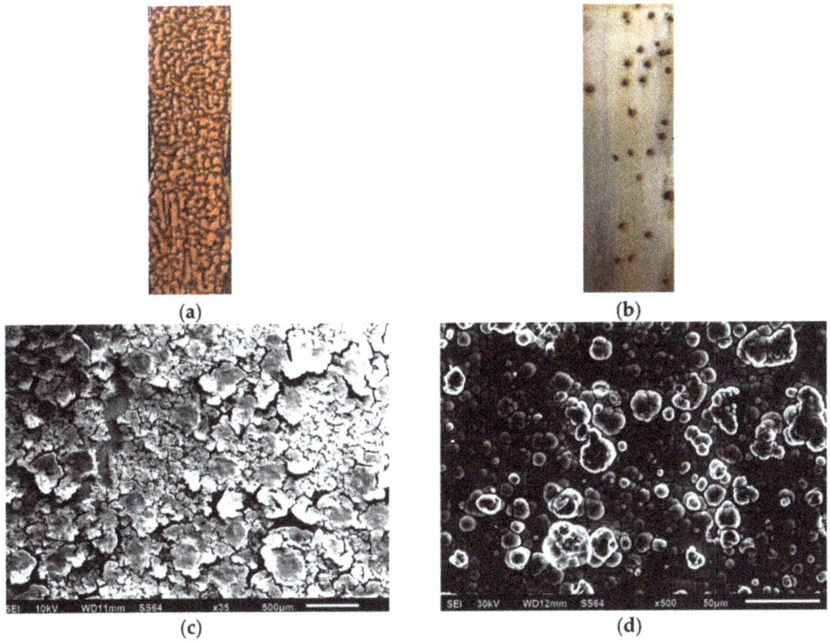

Figure 6. Macrograph of carbon steel surface (**a**) and Ni-P-rGO coating (**b**) after experiment; micrograph surface of carbon steel (**c**) and Ni-P-rGO coating (**d**) after experiment.

3.2. *Effect of Temperature on the Fouling Characteristics of Iron Bacteria on Ni-P-rGO Coating*

By maintaining other operating conditions, three groups of comparative experiments were performed by changing the temperature of the low temperature working fluid. The effect of the temperature on the biofouling characteristics of iron bacteria for three different temperatures is shown in Figure 7. As shown, the induction period existed in all the three experimental temperatures. When the temperature was 30 °C, the asymptotic value of the fouling resistance was the largest, while the induction period was the shortest. When the temperature was 35 °C, the asymptotic value of the fouling resistance was the smallest, while the induction period was the longest. When the temperature was 25 °C, the asymptotic value of the fouling resistance and the induction period were between the other two experiments. Figure 8 shows the variation of concentration of iron bacteria at different temperatures during the experiment. At 25 °C and 35 °C, the concentration of iron bacteria was almost always below that at 30 °C. With the increase in temperature, the Brownian motion was more intense, leading to an increase in the probability for iron bacteria to impact and adhere to the surface. Therefore, the time to reach the asymptotic value of fouling resistance was shortened at 35 °C. However, the activities of enzymes can also affect the reproduction and growth of bacteria [20]. At 30 °C, enzymes of the iron bacteria had more optimal activities than those at 25 °C and 35 °C. The iron bacteria with higher activity had a higher opportunity to adhere to the surface. Therefore, the induction period was the shortest and the asymptotic value of fouling resistance was the highest at 30 °C. In the initial stage of experiment, the number of iron bacteria at 30 °C was much more than those at 25 °C and 35 °C. But the iron bacteria reproduced and grew rapidly at 30 °C, and the coating corrosion was relatively serious. This led to increased exposed graphene and a fast consumption of the nutrients in the water. Hence, the concentration of iron bacteria declined rapidly after reaching a peak.

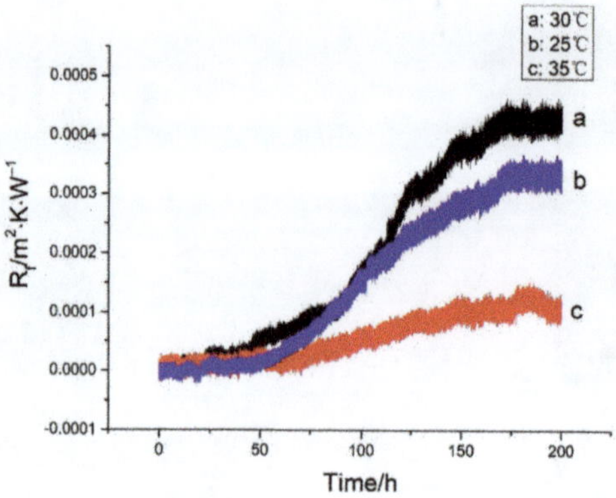

Figure 7. Effect of temperature on iron bacteria biofouling characteristics on the Ni-P-rGO coating.

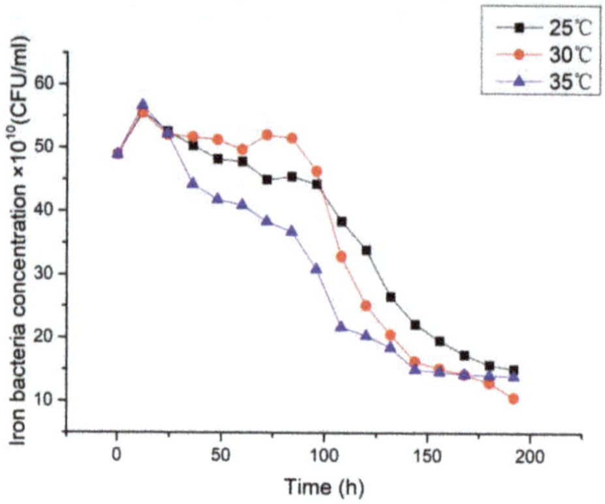

Figure 8. Variation of iron bacteria concentration at different temperatures in the low temperature working fluid.

3.3. *Effect of Flow Velocity on Fouling Characteristics of Iron Bacteria on Ni-P-rGO Coating*

By keeping other operating conditions unchanged, three groups of comparative experiments were performed by changing the flow velocity of the low temperature working fluid. The effect on characteristics of iron bacteria biofouling due to three different flow velocities is shown in Figure 9. As shown in the figure, with the increase in flow velocity, the induction period increased and the asymptotic value of fouling resistance decreased. In the fouling deposition stage, the growth and detachment of biofilm are simultaneous [3]. The accelerated flow velocity increased the shear force of the fluid on the coating surface, and the higher shear force peeled more growth biofilm. Therefore, the induction period increased with the increase in flow velocity. Furthermore, as the flow velocity increased, the thinner biofilm attached to the surface, therefore, the solid content was lower [4], and the asymptotic value of the fouling resistance decreased. However, higher flow velocity provided more oxygen and nutrients, which led to the shortening of time to reach the asymptotic value of fouling resistance at 0.35 m/s. Figure 10 shows the variation of iron bacteria concentration at different flow velocities during the experiment. As shown in Figure 10, the concentration of iron bacteria kept decreasing at 0.3 m/s and 0.35 m/s, but raised first and then declined at 0.25 m/s. Though higher velocity provided more oxygen and nutrients, the iron bacteria came into contact with graphene more frequently. Therefore, at 0.25 m/s, the reproduced and decayed iron bacteria reached a dynamic balance at the initial stage of the experiment. At 0.3 m/s and 0.35 m/s, decayed iron bacteria were more than the reproduced iron bacteria.

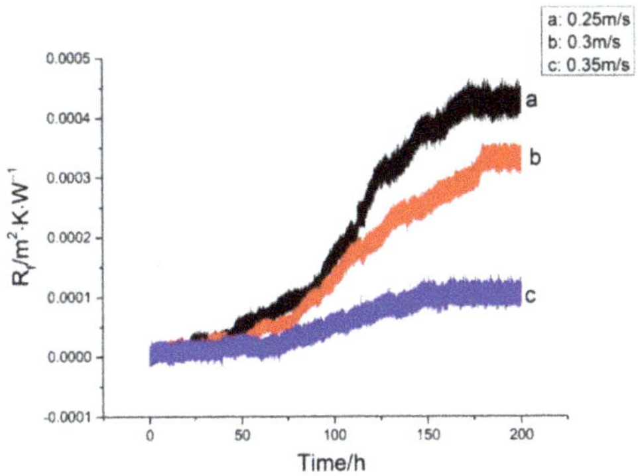

Figure 9. Effect of flow velocity on iron bacteria biofouling characteristics on Ni-P-rGO coating.

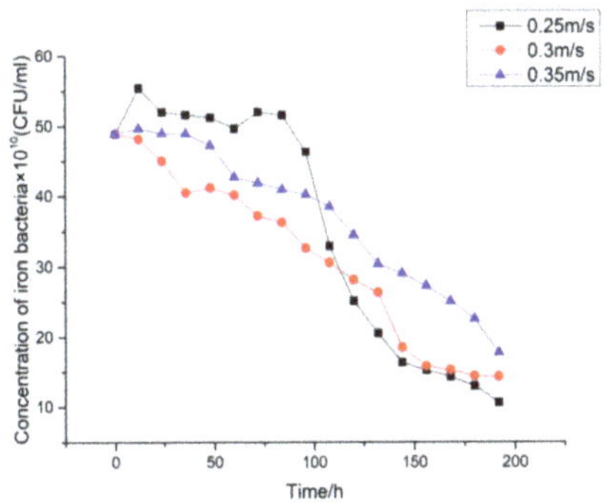

Figure 10. Variation of iron bacteria concentration at different flow velocities in the low temperature working fluid.

3.4. *Effect of Concentration on the Biofouling Characteristics of Iron Bacteria on Ni-P-rGO Coating*

By keeping other operating conditions unchanged, three groups of comparative experiments were conducted by changing the concentration of iron bacteria in the low temperature working fluid. The effect on the characteristics of iron bacteria biofouling for three different concentrations is shown in Figure 11. As shown in the figure, with the increase in iron bacteria concentration, the induction period decreased, while the asymptotic value of fouling thermal resistance increased. The initial conditions for the formation of the biofouling is the formation and adhesion of the biofilm, and the formation of the biofilm is related to the absorption of protein molecules and organic polymer [21]. As the concentration of iron bacteria increased, more protein molecules and organic polymers were secreted and the biofilm was formed quickly. Therefore, the induction period and time to reach the asymptotic value of fouling resistance were shortened. Furthermore, the asymptotic value of fouling resistance was increased. Figure 12 shows the variation in the different concentrations of iron bacteria during the experiments. At high concentrations of iron bacteria, there was more dramatic decay of iron bacteria. The reason for the phenomenon is that the increase in iron bacteria concentration led to more iron bacteria reproduction. However, when the nutrients in the water were depleted, the iron bacteria decayed dramatically.

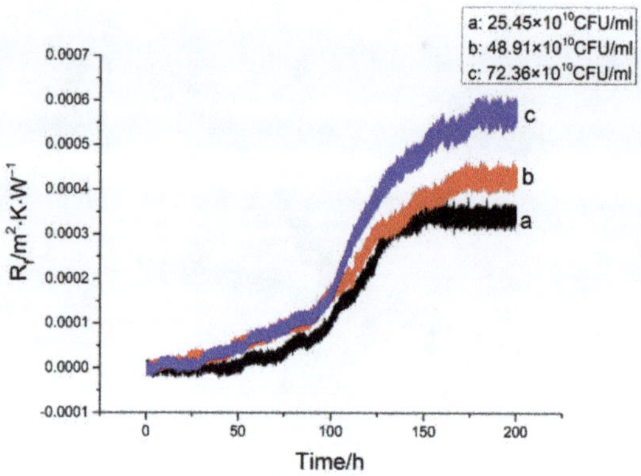

Figure 11. Effect of concentration on iron bacteria biofouling characteristics on Ni-P-rGO coating.

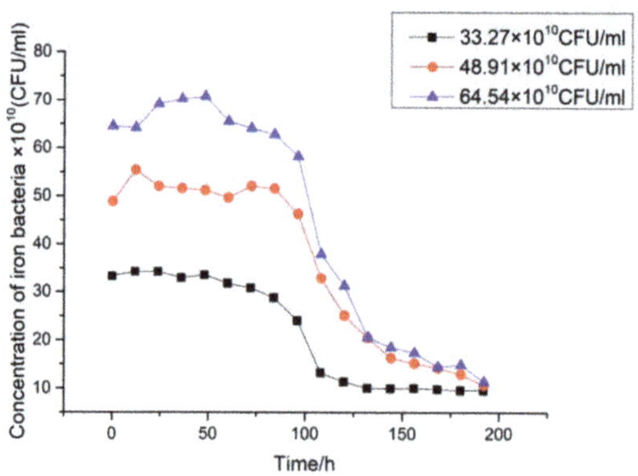

Figure 12. Variation of iron bacteria concentration at different initial concentrations in the low temperature working fluid.

4. Conclusions

[1] Compared to the carbon steel, which was used to prepare the heat exchange device as raw materials, the advanced materials with Ni-P-rGO coating had an important property that inhibited biofouling. They not only reduced the asymptotic value of the fouling resistance of iron bacteria fouling significantly, but also greatly increased the induction period and fouling growth time.

[2] The inhibition of advanced materials with Ni-P-rGO coating against iron bacteria biofouling at different temperatures was studied. The asymptotic value of fouling resistance was highest at 30 °C, and the induction period was the shortest at this temperature. The asymptotic value of fouling thermal resistance was lowest at 35 °C, and the induction period was longest at this temperature. The asymptotic value of fouling resistance and the induction period were between the former two at 25 °C.

[3] The inhibition of advanced materials with Ni-P-rGO coating against iron bacteria biofouling at different velocities was studied. With the increase in flow velocity, the asymptotic value of fouling resistance decreased, while the induction period increased.

[4] The inhibition of advanced materials with Ni-P-rGO coating against iron bacteria biofouling at a different initial concentration was studied. With an increase in the concentration of iron bacteria, the asymptotic value of fouling resistance increased, while the induction period decreased.

Author Contributions: Conceptualization, Z.X.; data curation, Z.X. and M.S.; formal analysis, Z.X., M.S., and Z.L.; investigation, M.S. and B.W.; methodology, Z.X. and M.S.; project administration, Z.X.; resources, Z.X.; supervision, Z.X., Z.L., and H.D.; validation, M.S., Z.L., B.W., and H.D.; writing—original draft preparation, M.S.; writing—review and editing, Z.X., Z.L., B.W., and H.D. All authors have read and agreed to the published version of the manuscript.

Funding: This research was funded by the National Natural Science Foundation of China, grant number 51976028.

Conflicts of Interest: The authors declare no conflict of interest.

References

1. Yang, S.R.; Xu, Z.M.; Sun, L.F. *Fouling and Counter Measures for Heat Transfer Equipment*, 2nd ed.; Beijing Science Press: Beijing, China, 2004; p. 15.
2. Li, B.; Logan, B.E. Bacterial adhesion to glass and metal-oxide surfaces. *Colloids Surf. B* **2004**, *36*, 81–90. [CrossRef] [PubMed]
3. Xu, Z.; Wang, J.; Jia, Y.; Geng, X.; Liu, Z. Experimental study on microbial fouling characteristics of the plate heat exchanger. *Appl. Therm. Eng.* **2016**, *108*, 150–157. [CrossRef]
4. Trueba, A.; García, S.; Otero, F.M.; Vega, L.M.; Madariaga, E. Influence of flow velocity on biofilm growth in a tubular heat exchanger-condenser cooled by seawater. *Biofouling* **2015**, *31*, 527–534. [CrossRef] [PubMed]
5. Chen, X.; Yang, Q.; Wu, R.; Zhang, N.; Li, N. Experimental Study of the Growth Characteristics of Microbial Fouling on Sewage Heat Exchanger Surface. *Appl. Therm. Eng.* **2018**, *128*, 426–433. [CrossRef]
6. Shao, W.; Zhao, Q. Effect of corrosion rate and surface energy of silver coatings on bacterial adhesion. *Colloids Surf. B* **2010**, *76*, 98–103. [CrossRef] [PubMed]
7. Xu, Z.M.; Wang, D.; Kong, L.W.; Liu, Z.D. Fouling characteristics of iron bacteria on the surface of electroless plating of Ni-Cu-P. *Chem. Ind. Eng. Prog.* **2017**, *2*, 47.
8. Liu, C.; Zhao, Q. Influence of surface-energy components of Ni-P-TiO2-PTFE nanocomposite coatings on bacterial adhesion. *Langmuir* **2011**, *27*, 9512–9519. [CrossRef] [PubMed]
9. Ji, D.X.; Fan, L.; Tao, L.; Sun, Y.J.; Li, M.G.; Yang, G.R.; Tran, T.Q.; Ramakrishna, S.; Guo, S.J. The Kirkendall Effect for Engineering Oxygen Vacancy of Hollow Co_3O_4 Nanoparticles toward High-Performance Portable Zinc-Air Batteries. *Angew. Chem. Int. Ed.* **2019**, *58*, 13840–13844. [CrossRef] [PubMed]
10. Tran, T.Q.; Lee JK, Y.; Chinnappan, A.; Jayathilaka, W.; Ji, D.X.; Kumar, V.V.; Ramakrishna, S. Strong, lightweight, and highly conductive CNT/Au/Cu wires from sputtering and electroplating methods. *J. Mater. Sci. Technol.* **2020**, *40*, 99–106. [CrossRef]
11. Wu, H.H.; Liu, F.; Gong, W.B.; Ye, F.Y.; Hao, L.F.; Jiang, J.B.; Han, S. Preparation of Ni–P–GO composite coatings and its mechanical properties. *Surf. Coat. Technol.* **2015**, *272*, 25–32. [CrossRef]
12. Qian, Y.; Zhou, T.F.; Jiang, Y.G.; Yan, X.; An, Z.L.; Wang, X.B.; Zhang, D.Y.; Ono, T. Preparation of graphene-enhanced nickel-phosphorus composite films by ultrasonic-assisted electroless plating. *Appl. Surf. Sci.* **2018**, *435*, 617–625.
13. Matjie, R.; Zhang, S.; Zhao, Q.; Mabuza, N.; Bunt, J.R. Tailored surface energy of stainless steel plate coupons to reduce the adhesion of aluminium silicate deposit. *Fuel* **2016**, *181*, 573–578. [CrossRef]
14. Shao, W.; Wu, J.M.; Liu, H.; Dong, G.H.; Wang, S.X.; Min, H.H.; Huang, M. Graphene oxide reinforced Ni-P coatings for bacterial adhesion inhibition. *RSC Adv.* **2016**, *6*, 46270. [CrossRef]
15. Lee, C.K.; Teng, C.L.; Tan, A.H.; Yang, C.Y.; Lee, S.L. Electroless Ni-P/Diamond/Graphene Composite Coatings and Characterization of their Wear and Corrosion Resistance in Sodium Chloride Solution. *Key Eng. Mater.* **2015**, *656–657*, 51–56. [CrossRef]
16. Chen, J.; Zheng, X.; Wang, H.; Zheng, W.T. Graphene oxide-Ag nanocomposite: In situ photochemical synthesis and application as a surface-enhanced Raman scattering substrate. *Thin Solid Films* **2011**, *520*, 179–185. [CrossRef]
17. Dinh, D.A.; Hui, K.S.; Hui, K.N.; Cho, Y.R.; Zhou, W.; Hong, X.T.; Chun, H. Green synthesis of high conductivity silver nanoparticle-reduced graphene oxide composite films. *Appl. Surf. Sci.* **2014**, *298*, 62–67. [CrossRef]
18. Wu, J.X.; Xu, H.; Zhang, J. Raman Spectroscopy of Graphene. *Acta Chim. Sin.* **2014**, *72*, 301–318. [CrossRef]
19. Akhavan, O.; Ghaderi, E. Toxicity of Graphene and Graphene Oxide Nanowalls against Bacteria. *ACS Nano* **2010**, *4*, 5731–5736. [CrossRef] [PubMed]

20. Garrett, T.R.; Bhakoo, M.; Zhang, Z.B. Bacterial adhesion and biofilms on surfaces. *Prog. Nat. Sci.* **2008**, *18*, 1049–1056. [CrossRef]
21. Hori, K.; Matsumoto, S. Bacterial adhesion: From mechanism to control. *Biochem. Eng. J.* **2010**, *48*, 424–434. [CrossRef]

MDPI
St. Alban-Anlage 66
4052 Basel
Switzerland
Tel. +41 61 683 77 34
Fax +41 61 302 89 18
www.mdpi.com

Materials Editorial Office
E-mail: materials@mdpi.com
www.mdpi.com/journal/materials